Agnes L. Sadlier

New lights or Life in Galway

a tale

Agnes L. Sadlier

New lights or Life in Galway
a tale

ISBN/EAN: 9783741131042

Manufactured in Europe, USA, Canada, Australia, Japa

Cover: Foto ©berggeist007 / pixelio.de

Manufactured and distributed by brebook publishing software
(www.brebook.com)

Agnes L. Sadlier

New lights or Life in Galway

NEW LIGHTS;

OR,

LIFE IN GALWAY.

A Tale.

BY MRS J. SADLIER,

Let coercion, that peace-maker, go hand in hand
With demure-eyed conversion, fit sister and brother;
And covering with prisons and churches the land,
All that won't go to one, we'll put into the other

Moore.

P. J. KENEDY AND SONS
PUBLISHERS TO THE HOLY APOSTOLIC SEE
3 AND 5 BARCLAY STREET NEW YORK
1905

DEDICATED

TO

THE PEOPLE OF IRELAND.

---◆---

To the faithful, and much-enduring people of Ireland, to those who still cling with undying love to the beautiful land of their birth, enduring all things rather than break asunder the tie which binds them to 'the Niobe of nations,' and to those who, like myself, have left the graves of our fathers, to seek a home beneath foreign skies—all alike bound together by the one glorious bond: our ancient, our time honored, our never-changing faith—to them do I dedicate this little work.

THE AUTHORESS.

MONTREAL,
    Feast of the Purification.

NEW LIGHTS;

or,

LIFE IN GALWAY.

"The good are better made by ill,
As odors crush'd are sweeter still."—Rogers.

FAR away in the extreme west of Ireland where
the waters of Lough Corrib reflect the changeful
hues of that ever-changing sky, there is a large,
straggling village running up along the bank of a
rivulet, from near the shore of the lake, for a distance of nearly two miles.   This village, which we
shall call Killany, though having in itself little to
interest the traveller, is still a desirable sojourn
for the summer months, ' while the grass is on the
fields and the blue is on the sky.'   The country
around is, indeed, beautiful, though somewhat wild
in its character, for the mountains of Connemara
stand like giant sentinels in the neighborhood,
receding from the inland view in many a grand
perspective.   Above the village, at a little distance,
the rivulet begins to assume the appearance of a

cascade, rushing down over the face of some pro-
jecting rocks with a force that sends the spray
a-dancing and glancing through the air. On and
on rolls the merry stream, dashing over its rough
channel, amid stones and fragments of rock, all the
way through the village—or rather beside it—till
it makes a way for itself through a limestone rock
to join the waters of the lake. The inhabitants of
Killany are for the most part poor, though there
are several families residing there who, as the saying is, hold their heads pretty high. Some years
ago there was a tolerably brisk trade carried on
across the lake, but these last miserable years have
considerably injured the village and its commerce.
Famine has been busy in the neighborhood, and
with it came its handmaid pestilence, and the misery of the people was great. It is true that Killany was not quite so severely scourged as some
other places, but still it had its full measure of
sorrow and suffering; and even now, when the
famine has exhausted its fury, there is still much
destitution existing in that locality. Here, as in
every other district of the south and west, ruin has
been busy amongst the farming classes, and many
a family has fallen, within the last few years, from
comfort and affluence and respectability, to want
and penury and utter destitution. The worst of
all is, that the distress is so general that those who
would gladly assist their neighbors, and often did,

have no longer the means for those who are not reduced to beggary, find it as much as they can do to maintain themselves, and 'keep the wolf from the door.'

Amongst the families who experienced the greatest reverses during those long dreary years, was one whose fall was a cruel blow to the poor of the vicinity. The father of the family, Bernard O'Daly, had been for many long years the strongest farmer about Killany. His farm was large, and well cultivated, his cattle of the best breed that could be procured, his barn and his haggard were plentifully filled year after year, and, in short, Bernard was always pointed out as a man particularly well to do in the world. In addition to these material blessings, Bernard O'Daly had a large family of sons and daughters, the like of whom were not to be found all the country over—his wife, it is true, was old, much older than her husband, and of broken health, but then she was surrounded by every comfort, and her periodical fits of sickness were of comparatively short duration, since her daughters had grown up to womanhood, for Kathleen and Bridget O'Daly, the two eldest, were the kindest and best of nurses—ay! and the best of housewives, too. There was a son older than they whose name was Cormac—a sedate, sober young man, who took upon himself the chief care of the farm—then next to Bridget

were two other sons, Daniel and Owen, and last
of all was a fair young creature, named Eveleen,
the pride and darling of the whole family. The
children of the O'Daly family had been well and
carefully brought up.   There was a very good
school in Killany, under the superintendence of
the priest, and though they had to go a distance of
a mile and a half to the village, yet the young
people had attended year after year, (each boy
and girl taking it in turn to stay at home to assist
their parents,) until they had acquired a very good
knowledge of the English language, which was to
them a foreign tongue, as their father and mother
spoke Irish for the most part.   When their children
grew up around them, all speaking English, then
the parents began to speak it too, and gradually
it became the language of the house, though not
without considerable grumbling on the part of the
old people, who still considered, and spoke of it as
'the strangers' tongue.'   But the gentle, modest
manners, and upright minds of the young O'Dalys
had not been acquired or formed solely in the
school house; they were, from their very infancy, regular attendants at the catechism, taught
and expounded every Sunday afternoon in the
parish chapel about a mile distant. They had
always been favorites with the priest, who was
well able to appreciate the worth of the family
and they, on their parts, listened with avidity to

every word of instruction, and loved Father O'
Driscoll better than any one else in the world,
their parents, perhaps, excepted.

While the world went well with the O'Dalys
they kept a servant girl and two men-servants
for caring the cattle and the two horses, and giving
a hand at the farm-work. Then there were some
ten or twelve laborers employed on the farm the
greater part of the year, and during the time of
putting in the crop, and again of taking it out and
gathering it in, there were many more employed,
both men and women. There was always plenty
of everything in Bernard O'Daly's, but never
anything like waste, for Honora, his wife, was
what is called a thrifty housewife, and brought up
her daughters in the same habits. Hospitality
was a virtue common to the whole family, and all
the country round could bear witness that theirs
was

"————————the door
That never was closed to the way-worn or poor."

But, as they said themselves, "it is a long lane
that has no turnin'," and though prosperity
attended the fortunes of the O'Dalys for many a
long year, yet the time came at last when they
were to have their trial, and to find everything
going against them, where before all had gone on
smoothly as the meadow-stream. The first of
their misfortunes was the fatal potato-blight of

45—then followed the death of cattle, and to make matters worse, a brother-in-law of Bernard's, named Lawrence O'Sullivan, absconded one fine night, leaving poor Bernard to pay forty pounds, for which he had gone security in the Provincial Bank in Galway. This was a fearful blow, for the money was not to be had, and so O'Daly was sued, an execution was granted, and every head of cattle he had was sold for the debt, together with his best horse. The clouds of adversity were gathering thickly around them, yet the cheerful piety of the family was proof against all, and when any of the neighbors set about condoling with them, the invariable answer was, " Well ! sure it's the will of God, and we have no right to complain—He gave us good things for this long time back, and it's our turn now to bear a little hardship. We're no better than others that's in want and misery on every side of us." One by one they had to part with their servants until all were gone, the daughters and sons remarking, by way of consoling their parents : " Well ! there wouldn't be any great use in keeping them now, for ourselves are more than able for all that's to be done." But still it was impossible not to feel, and feel deeply, the rapid though gradual destruction—the melting away, as it were, of all their goodly possessions, and though each individual tried to conceal it from the others, yet all were saddened and disheartened. For a while they

kept up the old appearance of respectability; as
long as the old clothes could be made to look any
way decent, there was no outward sign of poverty,
visible.   But alas! even the skilful industry of
Kathleen and Bridget could not keep things from
wearing out: they altered, and turned, and scoured,
and dyed, until the garments would bear no
more, and it was pitiful to see the consternation
of the whole family, when it was found that
"Cormac's best coat" or "Owen's buff vest" wasn't
worth "doing anything to," or that "father's brown
surtout" was "beginning to look very shabby."
There was no longer any means of replacing the
articles in question, and hence their decay was a
serious event to those who would fain have kept up
a decent appearance, at least "in the chapel on
Sunday," still hoping that better times would come
again.   Many a tear of sympathy was shed by their
neighbors, especially the poor, over the falling fortunes of the O'Dalys, and the change in their
personal appearance in chapel, or fair, or market,
drew forth many a heavy sigh.

"Och, then, Nelly dear," said one old woman
to another, as they sat together in a corner of the
chapel-yard after mass, one Sunday, "isn't it a
thousand pities to see the change that's comin'
over Barney O'Daly's family?"

"You may say that, Judy!" replied the other.
"I declare myself could cry for them, and sure

enough but it went to my heart this very day to
see the boys lookin' so shabby, an' the girls, too.
An' och! och! but it's they that never carried
their heads too high when they had full an' plenty
about them. The Lord comfort them this day,
and rise them out of the poverty again! I pray
that from my heart out!" and the good creature
raised her clasped hands and her tearful eyes to
heaven.

"Well indeed," said Judy, wiping her eyes,
" I'm a hungry woman this blessed day, and didn't
break my fast yet, an' God he knows I'd sooner go
without eatin' another day than see one of them
havin' an hour's hunger, for while they had it we
didn't want."

"True for you, *ahagur*. But what's that you
said about not havin' broken your fast? Was it
because you were goin' to communion?"

"Well! that was the raison," said Judy, "thanks
be to God for it, I did get communion, but," she
added, with a forced smile, " even if I wanted to take
my breakfast, I hadn't any to take, for I havn't a
mouthful of anything in the house that a body could
eat."

"Faix, then, it's lucky that you tould me,"
replied Nelly, "for, my dear, I've a beautiful
little dish of meal that I got last night from Nancy
McBreen, the priest's housekeeper—God bless
aim an' her botn, for it's hard to tell which of

them has the most feelin' for the poor ; an' so you'll jist come home an' take part of it with me—we'll have a good dinner at any rate."

" Well, but, Nelly *astore !*" pleaded poor Judy, " What will you do when the meal's done, an' you havin' the two little grandchildren with you ? Many thanks to you, but I'll not go ! I have only myself, and I'll get a mouthful somewhere that'll keep the life in me."

"Nonsense, woman !" cried Nelly, almost angrily, " do you think I'd let an ould neighbor go off to look for a chance bit, an' me havin' something at home ? Jist get up now and come along home with me—never you mind what I'll do when the meal's out—we'll lave that to God."

" Well ! well ! I see there's no use in excusin' myself," said Judy, standing up, as her friend had already done, " sure I know that God won't let you suffer for dividin' your little bit with one poorer than yourself." So saying, she drew the hood of her old cloak over her head, and the two old women hobbled away.

Meanwhile there was a dialogue of a different character going on in another corner, under the shade of a large old sycamore tree. The speakers were a young man and an old one : the former clad in a faded-looking blue coat with brass buttons, and pantaloons of drab cloth considerably the worse for wear, and the latter in a dark brown

2

surtout, with knee-breeches of gray corduroy, and a broad-brimmed hat that had once been a good beaver. The old man stood leaning on a stick with hands clasped, and was speaking in an earnest tone. It seemed that both were waiting for the appearance of some person. "I couldn't bring myself to speak to the master himself about it, Cormac!" said he, "but, of course, it will just do as well to speak to the priest. At any rate, there's no use in letting it run any longer, for every day will make the matter worse."

"Well, now's your time, father," said the young man in a low voice, "for here's Father O'Driscoll now. So while you're talking together, I'll just go over there to Larry Doolan that I see at the chapel-door, and have a talk with him about America." Larry had but recently returned from the United States.

"Good morning, Bernard!" said the priest, as he approached the old man, who was no other than our friend Bernard O'Daly. "How is all with you to-day? It is something new to see you at chapel alone!"

"A good morning kindly to your reverence!" and the old man touched his hat respectfully. "We're all middling well as to the health, thanks to you for asking, and sure the whole family was at mass, thanks be to God, except the old woman herself that's not very strong, you know, an' little

Eveleen that staid at home to keep her company. The boys and girls hurried off home as soon as mass was over, only Cormac that staid to be back with me. He's just gone over there to speak to Larry Doolan."

"Indeed!" said the priest, who was a man of some forty years or thereabouts, with a pale, intellectual countenance, and dark, thoughtful-looking eyes. "Has he some thoughts of America then?"

"Well! I don't know, your reverence," replied the old man hesitatingly, "times are bad here, and we havn't the same way of doing that we had. Cormac thinks—poor fellow! that if he was in America he could do something to help us, and—'

"And so he might, Bernard! and I would strongly advise him to go, and Daniel, too—it is the very best thing they could do."

The poor father fetched a deep sigh as he answered—"I believe it is, your reverence,—I believe it. But sure I know you havn't broke your fast yet after sayin' mass, so I can't be keepin' you standing here. I just wanted—ahem! to ask a little favor from your reverence."

"Well!" said the priest, in a kind tone, "I only hope that you are going to ask something that can do, for it would go hard indeed with me to refuse you the first favor you ever asked of me."

"I know that, sir, I do indeed, and that's the raison why I came to trouble your reverence at

this time. It's about the school-money, Father O'
Driscoll, that's due the master beyond for my
little girl and boy. I'd be very willin' to pay it if
I could, your reverence, but"—the old man stopped—coughed—his thin cheek grew somewhat
redder—" but—I may as well tell the truth at
once—I can't raise the money, do what I will. So
I just came to ask your reverence to interfere with
the master for me—if God sends me the means
again I'll pay him, an' that's all I can do."

The priest was silent for a moment—walked a
step or two away—then turned back, took out his
snuff-box and took a pinch, then handed it to
Bernard. This gave him an excuse for using his
handkerchief.

"Sad times these, Bernard! sad, sad times"—
the handkerchief was again used. "Well, my
friend—my good, my long-tried friend, so you
were unwilling to speak to me on this business—
ah! Bernard O'Daly! I think you should have
known me better!—well! no matter. I'll settle
it with McEgan—send the children to school as
usual."

"I will, your reverence, an' many thanks to
you—that is, we'll send them if we can keep
clothes on them. Ahem! ahem!—well! good
morning to you, sir, an' my blessin' be with you
'us day an' every day you rise."

"Good bye, Bernard!" said Father O'Driscoll,

warmly shaking the old man's hand—"tell Mrs
O'Daly I'll be over one of these evenings to see
her."

"Do, sir, an' God bless you, for we want to
have your opinion about the boys, an' some other
little matters that poor Honora has on her mind

Father O'Driscoll made a sign to the boy who
was walking his horse up and down. but before he
got his foot in the stirrup, he was stopped by a
pale, delicate-looking woman with a young child in
her arms: "Could I have a word with your
reverence?"

"Well, my good woman—oh! you're Katty
Boyce—well! Katty—what's wrong with you?"

Before the woman could answer, a man standing
by exclaimed: "Don't mind her, your reverence,
she gets soup from the Bible-readers—she's not
to be trusted, Father O'Driscoll."

"Well!" said the priest calmly, "I must hear
what she has to say?—is it true, Katty, that you
take ' the soup?' "

"Och, then, it is, your reverence—God help
me! it's true enough, an' that's just what I was
comin' to spake to you about, sir," throwing a
reproachful look at the informer: "people shouldn't
holla till they're out o' the wood—may be your
self, Tom Hynes, might have to call at the
soup house before all's over."

'Well, well, Katty," said the priest, "never mind Tom, but tell me what is the matter with you."

"I will, your reverence. You know, sir, ever since our Micky died, myself an' and the children's in the heighth o' distress—I needn't tell you that, for many's the time you relieved us—well! about three weeks ago Mrs. Perkins—you know her, your reverence—the lady that goes around with the tracts—well! she persuaded me to go an' apply for some o' the soup an' bread' an' when I said that I wouldn't go on any account for fear I'd have to go to church, or get my name down as a Prodestan', oh! she was as sweet as sugar, an' tould me that I mightn't be the laste afeard o' that, for that she'd put in a good word for me, that I wouldn't be asked any questions at all about my religion—'for,' says she, ' my poor woman! I *do* feel very much for you—indeed I do!' so, sir, to make a long story short, I went every day with my can an' got some soup an' a loaf o' bread, an' for a week or so there wasn't a word said about religion, but last Saturday week, Mr. O'Flanagan that gives out the soup began to me in style, an' he said if I didn't let my name be put down in the book as a Prodestan', I might go far enough before he'd be servin' me every day. Well! sir, I tould him plump an' plain that I wouldn't, an' so he bid me be off, an' never to shew my face again unless I'd do what they wanted. I staid away two or three days, and tried to gather a bit among the neighbors, but ochone! they hadn.'t it for themselves, the craturs! let alone for another, an' the weeny things were cryin' with the dint of hunger, an' myself didn't know what to do. I prayed to God to keep me from the temptation, an' to give me some way to keep us all from starvin', but no relief came, an' after we were a whole day an' night without tastin' bit, bite, or sup. I got a'most crazed listenin' to the pitiful cries o' the children, au' off I runs again to the soup-house: 'Well!' says O' Flanagan, says he, 'you're back again, are you?' 'Yes,' says I, 'I'm comin' to ask charity from you again.' 'Ha! ha!' says he, 'you see you can't do without us after all. I suppose I'm to put down your name now!' an he brings out a big book, an' sure enough when I looked at it I began a tremblin' all over. 'Here now,' says he, dippin' his pen in the ink bottle, 'what's this your name is?' 'I was in hopes, sir,' says I, 'that you'd give me a little help for this day, without askin' me to get down my name—do! an' God bless you!' 'Not as much as would fall from our finger,' says he back to me, an' then he began to look very angry, an' says he, 'Get you gone, you ignorant slave of'—something—I don't remember what the other long word was—'never darken this door again, you may starve and die like a pig, for you're no better.' With that, your reverence, up comes a smooth-faced, well-spoken gentleman with a black coat an' a white cravat, an' he says to O' Flanagan, 'Don't be so short with the poor woman —you converts from Romanism are over-warm at times'—then turnin' round to myself he says to me, 'Since you have such an objection to have your name registered as a Protestant, my good woman, we must not be too hard with you—you are well recommended to us by that good lady Mrs. Perkins, so you can have whatever provisions you want, without being put on the books—which is, I assure you, a great favor!—only send your chil dren to our school—that will do you or them no harm, nor will it at all affect your religion!' 'Well, but, sir,' says I, thinkin' to get off, 'the children havn't a stitch on them.' 'No matter,' says he, 'no matter for that, you know we give good, comfortable clothes to all the children attending our schools.' When I seen the new tack they were on, I thought I'd jist come an' ax your advice, Father O'Driscoll, an' that I'd do the best I could till then."

"Well! and what did you say to this last proposal?" inquired the priest.

"Why! I said, sir," said the woman hesitatingly, "that I couldn't consent to that at all, an' that I'd sooner they'd put my name in the book than to send the children to a Prodestan' school—for that

I wouldn't put them in danger for the world—I'd
sooner we'd all die of hunger than that."

"So you allowed them to put down your own
name as a Protestant?"

"I did, your reverence—the Lord forgive me!
—because I said to myself that so long as they
didn't ax me to go to their church, it didn't make so
great a difference, an' that so long as God an' your
reverence an' all good Christians would know me
for a Catholic, I mightn't care much about them
havin' my name in their book."

Father O'Driscoll smiled at the poor woman's
logic. "Well! but what did they say when you
refused to send your children?"

"O'Flanagan's face grew as red as a turkey's
head, your reverence, an' he was for orderin' me
away altogether, but the smooth-spoken man said
to let me alone in my own way, an' sure enough I
seen him winkin' at O'Flanagan, an' so they
entered my name in the big book—the Lord in
heaven forgive me for that same!" and she crossed
herself devoutly, " an' I got my soup and my bread
regular these three days back. So that's jist what
I wanted to spake to your reverence about."

"Well! my poor woman," said the priest, who
had listened to her long story with one foot in the
stirrup and the other on the ground, "you have
done very wrong in having anything to do with
these people, but I am glad to find that you refused

to send your children to the Protestant school.
They would have been better pleased to have had
the children even than your name, for they have a
greater chance of succeeding with them than
with the parents. Keep your little ones from
all communication with them," he added, raising
his voice so as to be heard by all around, "as for
yourself, God direct us for the best!" he sighed
deeply—"we must endeavor to do something for
you, so that you may not be obliged to solicit
charity again from those tempters."

"Ah! then, that's the droll charity that they'll
give," said a stout, chubby-faced man coming forward—"sure they wouldn't give the weight of a
pin to save every man, woman and child about
Killany from starvation, if it wasn't that they're
trying to buy us up. Stand out o' my way there,
you Peery Boyce, till I spake to his reverence.
Father O'Driscoll, sir, I'm in want of a woman—
or at least the woman that owns me is—to spin
some wool, an' as poor Katty here is so badly off,
I'll take her out o' the hands o' the Philistines.—
Just come over in the mornin', Katty, and begin
your work, and you can tell the children to come over
after you till they get their breakfast—let you and
our Nanny agree between you about the payment."

"God bless you, Phillip—God bless you and
yours!" said the priest, fervently, as he mounted
his horse—"may He repay you an hundred fold!—

you are doing me a great kindness, too, by giving
this poor creature some employment—now, Katty !
you see how God has raised up a friend for you in
your greatest need—return thanks to him for this
new favor !"

Katty was now laughing and sobbing alternately.
"Sure I know that, your reverence, I know that !
My blessin' an' the blessin' of God be about you
both ! Ah ! ha ! now I can go an' order them to
take my name off o' the cursed book—oh ! glory
be to God this day ! I'll be over in the mornin',
Mr. McGuire, as soon as the day breaks—if God
spares me till then !"

The crowd now separated to make way for the
priest to pass, Phil Maguire following after mounted
on a white pony, and as the good man rode slowly
along in the wake of the priest, he was greeted by
a cheer such as only the warm-hearted Celt can
give. The story was quickly whispered around,
and before honest Phil could reach the gate, he
heard his name pronounced on every side, accompanied with abundant blessings. Phil began to
feel quite ashamed, and muttering, "what a fuss
they make about nothing !" he pushed his little
nag to her utmost speed, and at last escaped from
the chapel-yard, much to his own satisfaction, and
to the no small amusement of the priest, who had
been a silent observer of the scene.

"So you have got away at last, Phil ?" said

Father O'Driscoll, as they rode along side by side, for their road lay in the same direction.

"I have, your reverence," and Phil wiped the perspiration off his flaming countenance, "an' if it wasn't a job to make my way through, never say my name's Phil Maguire. Well! the craturs! if they don't beat the world wide for gratitude!—it's a thousand pities to see them as they are!"

"You may well say that, my good friend!" observed the priest, "but you know it is written that those whom God loves he chastises—that is our only consolation. Oh! that is your way home—well! God bless you, Phil."

CHAPTER II.

"Puck found it handier to commence
With a certain share of impudence,
Which passes one off as learned and clever
Beyond all other degrees whatever."

MOORE, *Song of Old Puck.*

WHEN Bernard O'Daly was rejoined by his son, they set out for their home, and on the way Cormac told his father that he had made up his mind to go to America, provided his parents gave their consent. "You know, father dear," said he, "that from the present state of this unfortunate country, we have none of us any other prospect than that of endless misery, and if God spares me to reach America, I hope I'll be able to do something for you all. That's the main object I have in view."

The old man sighed heavily, and for a moment he made no answer. When he did speak, his voice was low and tremulous. "I know, Cormac —I know you mean well—an' I can't blame you for wantin' to go, because I see plainly that there's nothing to be done here—but then—ah! my son, it will be a sore, sore crush to your poor mother—

not to speak of myself—och, it will, it will, indeed,
Cormac!"

"Well, but, father," resumed Cormac, gulping
down as well as he could his own strong emotion,
"you know it's the very best thing I can do—it
may be the means of taking the whole of us out
of poverty—and, besides, I may be only a year or
two away till I ca. send for you all—perhaps
come home for you! Think of that, father dear!"

"I do think of it, Cormac, but God only knows
what might happen to us all in two years, or even
one—you might find us in our could graves, my
son, when you would come. But, sure, sure, if
it's the will of God for you to go, it wouldn't be
us that would say against you. An' after all, I
think it is His will, for Father O'Driscoll advises
me to let you go—ay! an' Daniel too. Well!
well! there's many a fine family scattered over
the world in these times, an' sure we must expect
our share of what's goin'—good morrow to you,
Phil."

"Good morrow kindly, Bernard!" returned our
acquaintance, Phil Maguire, reining in his white
nag; "how's all with you the day?"

"Well, thank God an' you. I hope you've the
same story to tell!"

"Why, for the matter o' that, Bernard, we're
all in good health, an' as long as God spares us
enough to eat, we'd be very ungrateful to complain.

Why, Cormac, my man! what's the matter with you? I declare to my heart your face is nearly as long an' as sour as if you were one o' the Bible readers. Arrah! hould up your head like a man, it's newens for you to be down-hearted."

"Well, if you knew but all, Phil!" said Bernard, confidentially lowering his voice, and sidling up close to the nag—"it's no wonder he'd be down-hearted now—and we too—sure there's black trouble on us!"

"Why, is there anything new?" said Phil, earnestly.

"Listen here, Phil!" Maguire bent his head to listen. "Sure, Cormac is goin' to leave us—ay! indeed is he—goin' away to America, poor fellow!"

"Pho! pho!" said Phil, evidently much relieved, "if that's all, I don't pity you much. Why, man, myself thought by the hummin' and hawin' you had—not to speak of Cormac's pitiful-lookin' face—that some o' the young ones had been wheedled away by the Jumpers! My sowl to glory, O'Daly, but it's glad I am to hear what you tell me. An' when do you intend to start, Cormac, aroon?"

"That depends on my father and mother, Phil. It will not take long to prepare, I suppose, but—between you and me, Phil—" Here it was Cormac's turn to look round, and lower his voice "between you and me, it will be no easy matter for us to make up enough to pay my passage. That's what grieves me most, for I know it will leave them all bare and naked for many a day to come."

Phil was suddenly taken with a bad fit of coughing, which he did not get over for some minutes, and by that time he had reached the foot of his own lane. "Why, then, bad manners to this cough," said he, clearing his voice, "it's always at the wrong time it comes on me, so it is. Well! God be with you both till I see you again—an' above all, be sure an' keep your hearts—all's not lost that's in danger—d'ye mind me now?" And without waiting for an answer, honest Phil turned down the lane, and rode leisurely home. When he had put up his nag, Phil went into the house, and whilst waiting for his dinner, began to give his wife an account of the morning's work he had been doing. Having first ascertained that Nanny was somewhat recovered from the sudden sickness that had kept her from going to mass, he went on with his story, and when it was ended he called upon Nanny to "rejoice and be glad," for that they had the means of saving poor Katty and her children from the jaws of temptation.

"Humph!" says Nanny, who was far from sharing her husband's liberality, "It's great cause for rejoicin' I'm sure—I declare it's an estate you

ought to have Phil Maguire—nothing less is any
use to you —an' then you could gather all the beggars in the county round you."

"An' I'm blest but I would, Nanny!" rejoined
the husband, "hut, tut, woman, don't be so hard—
sure, when God is so good to us as to give us full
an' plenty, isn't it the least we can do to divide it
with them that's in want? Bless my soul, Nanny!
won't it be all here after us—we can take none of
it to the grave with us!"

"Well, well, there's no use, I know, in talkin'
to you, for you're ever an' always the same—but
come over here an' take your dinner. Myself
doesn't much care for that Katty Boyce—there's
many a creature in the parish that's as badly off
as she is—ay, troth is there, hundreds o' them—
that wouldn't go near the Jumpers—don't tell me
about her bein' in want, that's no excuse, Phil!—
no excuse in the world—she's not the thing, I tell
you!"

"Come, now, Nanny," said Phil, coaxingly,
"don't be too severe on the poor creature—neither
you nor I knows what we'd do if we were starvin'
with the hunger, an' listenin' to the children cryin'
for what we hadn't to give them—there's no danger of that with us, any how, Nanny, because we
hain't them to cry—but sure, may be we'd do
that an' worse—well, no, we couldn't do worse,
let us do as we would—but at any rate, Nanny

dear, it's hard to stand hunger—bedad it is so
An' them villains o' the world knows that well!"

"Nanny," said Phil, after a while's silence, "I
hope you'll stir yourself an' get that wool spun as
fast as you can—if you'll only have me two or
three pairs of good long stockings knitted in the
course of a fortnight or so, I'll buy you an elegant
new shawl when they're finished."

Nanny stopped short in her work —she was.
washing the dinner dishes—and fixed her keen gray
eyes on her husband. "Humph!" it was her
favorite interjection, "Humph! there's something
else in the wind now," said she at length, "and
you may just as well tell me what it is at onst."

"Indeed an' I will then, Nanny, for I don't want
to keep it a secret from you. Poor Cormac
O'Daly is goin' to America very soon!"

"Well, an' what's that to us?" said Nanny,
gruffly.

"It's only this," returned Phil, resolutely, "that
I want to give him something to keep his legs
warm when he's away next winter far from his
mother and sisters, an' where he won't have Nan
ny Maguire's nimble fingers to knit him a pair o
two of stockings. They're all strangers where the
poor boy's goin', an' you an' I knows him a long,
long time, Nanny, dear—"

"Ay, that's always the way with you, Phil,"
snapped Nanny again, "when you're wantin' to

some round me any way, you can give a good rub
of the blarney—Nanny Maguire's nimble fingers,
as you call them, have something else to do besides knitting stockings for Master Cormac O'Daly.
Let his mother and his dandy sisters knit for him."

" So you won't do it, Nanny ?"

" No, nor the sorra a stitch, Phil."

" Well, well—it can't be helped," said Phil, affecting a tone of disappointment, knowing well, all
the time, what would happen, for he could wind
up Nanny, with all her sharpness, just as easily as
he wound up his huge silver watch.

No more was said on the subject, but the very
next day the nimble fingers were busily at work
on what Phil well knew to be Cormac's stockings,
though he affected to take no notice of what was
going on.

Early in the morning came Katty Boyce, pale
as ever, with her still paler infant on her arm—the
poor child was more than two years old, yet still
helpless and feeble from the total want of nourishment. While his mother worked, he sat on the
floor at her feet playing with one little thing or
another, and looking much better than usual, thanks
to Nanny Maguire's kindness. The other two
children came about breakfast time, and having
got their little stomachs well filled with good oat-
meal-porritch and fresh milk, they returned home

to fetch their books, as Phil declared they must go to school that very day.

"What books have they, Katty?" said Phil, turning into the house after seeing them off.

"Well, indeed, myself doesn't know, Misther Maguire. They're two little books that Father O'Driscoll gave them when they were goin' to his school."

"Oh, very well, Katty, that's the very thing, for when the priest gave them to them they're sure to be the right sort. I'll go out now, Nanny, honey, to see how the men's gettin' on in the field abroad, an' when the children comes you can give me a call, an' I'll go down with them to the school-house."

But though Phil staid in the field till dinner-time, there was no call given, for the children did not return, and great was their mother's fear lest they had met with some mischance. Phil came in at last, to his dinner, and his first question was about Katty Boyce's children. No one could tell anything of them, and Phil began to ruminate. After sitting silent a few minutes, with his eyes fixed on the flickering turf fire, the muscles of his face began to work, his ruddy cheek waxed ruddier still, and at last he clenched his fist and started to his feet, evidently under some sudden inspiration : "Ah, the kidnappers ! the thievin' villains !" he exclaimed, in a tone that made the women start, while the men at the table dropped their knives and forks, and looked round in amazement at their master.

"Why, what in the world wide ails you, Phil?" cried Nanny, anxiously.

"What ails me!" shouted Phil, who was hurrying to and fro across the kitchen in a way that excited no small fear for his senses. "Isn't it enough to make any man mad to see them white-livered dogs — the Jumpers — prowlin' about like wild beasts, watchin' for their chance to pounce on poor little innocent children, and miserable starvin' creatures, an' draggin' them away to their den— now, I'll tell you what it is," he said, in a somewhat calmer tone, and stopping short in his walk in front of Katty Boyce, "the kidnappers have caught them children of yours, as sure as you're sittin' there."

"The Lord save us, Misther Maguire—do you think so?"

"Think so!—I tell you I'm sure of it."

"An' so you may, masther," said one of the men, Patsey Rearns by name, "for I've seen them myself as often as I've fingers an' toes on me, gettin' into discoorse with the children along the road goin' to school, an' tryin' to palaver them."

"But sure they wouldn't be so bould as to take the little creatures away with them that way in the middle o' day-light. Wouldn't they know that their people would be lookin' after them ?"

"Fiddle-de-dee !" cried Phil, contemptuously. "You know nothing at all about them, Nanny ; I tell you they'd do anything—anything in the world, a'most, to get the name of a convert—man, woman, or child—good, bad, or indifferent—it's all the same to them, only get old or young away from Popery. D'y hear me this, boys ?" he said, addressing the men, " wouldn't it be a good deed to go an' see what has become of these children, just for the fun of it—let alone the charity that it will be—eh now—what do you think ? which of you'll come with me--I only want one or two ?"

"Indeed, I'll go, in a thousand welcomes !" said Patsey. And the others having all answered in like manner,

"Be easy, now," said Phil, " I'll just take Patsey an' Brian here—let the rest of you go back to your work—but sit a little while first, after your dinner. By the tar o' war we'll have some fun this very day, or I'm not Phil Maguire of Ballyhasel."

Nanny remonstrated, remarking, that it was " easy to get into trouble, but not so easy to get out of it;" and Katty, her eyes swimming in tears, begged him not to put himself in any danger on account of her children.

"Danger !" said Phil; " oh, the sorra much danger there's in it—them lads knows Phil Maguire of old, an' they'll get rid of me as easy as they can. Put on your coats, boys, an' come along. Don't be afeard, Nanny, there will be nothing wrong, I warrant you."

"Well! God grant it!" sighed Nanny, making a virtue of necessity.

It was but a few hundred yards across the fields to the pleasant knoll whereon was seated the great Protestant school-house of the district—to wit, a long, low building, with a slated roof, "white-washed wall and nicely-sanded floor." Phil walked right up to the open door—rested his hand on either side against the door-cheeks and popped in his head: (having stationed his companions outside in waiting against any emergency.)

"The top o' the mornin' to you, Mister Jenkinson!" The long-faced functionary nodded, or rather bent his head. "I want them two little boys of Katty Boyce's that you have in there," continued Phil.

"I know not who you mean, my good fellow!" said the grave teacher in a grave voice. "Wherefore do you come at an unseasonable hour to interrupt the peace of the school?"

"But I know myself who I mean, my good fellow!—if that's your word—an' you knew as well as I do, an' if you don't give me out the children, I'll ' interrupt the pace' in earnest for you Out with them here in one minnit!—no hagger taug

gerin' there in the corner !" seeing the cadaverous-
looking pedagogue whispering with an elderly female
at the other end of the school.

" Man !" said Jenkinson, coming slowly forward,
" man, I know you not—by what authority do you
claim these children—that is, if they are here—
which I say not that they are ?"

" I'll soon see whether they are or not," returned
Phil, and raising his voice to the highest pitch, he
called out—" Jemmy and Terry Boyce, where are
you, children ?"

" Here, sir !" and " here, sir !" answered two
faint voices from the lower end of the room, and
two little pale faces were raised over the heads of
the others.

" Come out here, then !" said Phil, " your mother
wants you at our house." The little boys were
running to the door, when the lady—it was that
excellent woman Mrs. Perkins—took hold of one,
and Jenkinson grabbed at the other, like a shark
in danger of losing his prey.

They shall not go from here !" said Mrs. Perkins
in a tone of authority, " I myself induced them to
come in this morning when I met them on the
road—and they shall not go hence unless their
mother comes for them."

" Arrah! by your leave, ma'am !'" said Phil,
with a comical smile, at the same time beckoning
over his shoulder to Patsey and Brian, " sure such

a nice, genteel lady as you wouldn't keep any poor woman out of her own children?—you would—would you?—well, here, boys, come in an' take the little ones."

Jenkinson shrank back, and let go his prey, when the brawny arm of Patsey was laid on the boy's shoulder; but Mrs. Perkins waxed strong in her womanly prerogative, well knowing that no violent hand would be laid on her. "I tell you, men! that you shall not have the child—agents of the devil you are, and I cannot consent to give up the poor boy to your machinations. Go back and tell your priest who sent you here, that the child is in safe keeping until his mother comes to claim him."

"You're under a triflin' mistake, ma'am," said Phil, coolly; "the priest knows no more about what I'm doin' than the man in the moon. But if it goes to that, what business had you to wheedle the boys in here?—tell me that now!"

"I found them idling along the way-side," returned Mrs. Perkins, "and so I prevailed upon them to come in and listen to the words of wisdom. My heart yearned over the innocent children, and the spirit within urged me to draw them into the right path—the path of solid learning and Scriptural knowledge!—would that I could win you too, my dear good man! to enter upon the way of heavenly light!"

"Mai y thanks to you, ma'am," said Phil, casting
a sly glance at Brian and Patsey, " but as I happen
to be in the way you speak of already, I don't
think you'll make anything of me—so the bait
won't take, an' you needn't bother yourself throwin'
it out—I think your black-avised friend there can
tell you that, though he pretended not to know me.
Come, boys, take the children an' let us go; we're
only losin' our time here."

"Maguire !" said Jenkinson, when he saw the
two boys walking away with Patsey and Brian,
"Maguire, I'll make you repent of this job—and
that before you are much older !"

"Ah then, d'ye tell me so ?—an' how will you do
it, mister Bible-reader ?"

" Remember your landlord is not a Papist —he
shall hear of your insolence before the sun goes
down. I warn you I will do my utmost against
you, and so will Mrs. Perkins, who is a daily visitor
at the Hall !"

"An' to the devil I pitch your tale-bearing an'
your warnin' to boot, Mister Jenkinson !—I'm
like the miller of Dee, my lad." And raising
his voice he sang :

> "I pay my rent on quarter-day
> My wife and I agree,
> I care for nobody—no, not i,
> Nor nobody cares for me."

Then snapping his fingers to make his meaning

more clearly understood, he said: "I don't care
that for Mr. Ousely, and Mrs. Perkins and Mr.
Jenkinson all put together." Then putting in his
head again, he cast a pitying glance over the long
lines of miserable, hungry-looking little faces, and
then at their comfortable though coarse clothing :
" May the Lord look down on you all, this blessed
day—poor, desolate cratures that you are—most
of you without father or mother—better for you,
a thousand times, poor sorrowful things, that you
were lyin' under the turf with them that own'd
you, than to be robbed of the religion that would
bring you to heaven !—och ! och! but it's the sorrowful sight," said poor Phil, wiping away a tear
that would find its way to his cheek, "it makes
my heart black an' sore to see so many of you
gathered in there that ought to be Catholics—the
Lord pity you for poor children !"

"Be off instantly," said Jenkinson, "or I'll
horse-whip you!"

"An' if you'd only try it," retorted Phil, " you'd
find that two could play at that game. Now mind
what I tell you, my honest man—I beg your
pardon for calling you out of your name! Jist
let these children alone—that's my advice to you
—you know very well that I'm a man of my
word, an' I promise you if ever you lay hold of
them again, when you get them out alone, as sure
as anything I'll take you neck and heels, and put

you in the first bog-hole I meet. Now that's my
parting word to you—good mornin', ma'am,"
bowing very low to Mrs. Perkins, "when you're
makir' your report up at the Hall, will you jist
say for me that you're all runnin' a wild-goose
chase—beatin' the wind, ma'am, an' you know
that's mighty unprofitable work."

Phil hastened on after his aids, but they had
got the start of him, and were already out of
sight. He took a near-cut across the fields, and got
home in time to introduce the children. Coming
up panting and blowing just as they got to the
door, he began to holla at the top of his voice,
"Now, Nanny!—now, Katty Boyce!—what did I
tell you—eh?" At the first sound of his voice,
accompanied by the loud laughter of Patsey and
Brian, out ran Nanny, (her knitting in her hand,)
Katty Boyce, and Pincher, the big watch-dog—
the latter quite loud in his gratulation.

"Now, Katty!" repeated Phil, unbuttoning his
coat, and shaking it back from off his heated
chest, "there they are, my woman! safe an' sound
for you."

"The Lord bless you, sir!" said Katty. "But
where did you find them at all, at all—the ramblin'
little villains?"

"Just where I expected to find them—there
abroad in Jenkinson's school."

"Well! but what took them there?" cried

Nanny—"I'm sure they didn't go in of them-selves."

Katty was running at the children with uplifted hand, and the boys began to blubber, but Phil interposed his substantial person between them and their angry parent. "There's no use in beatin' them, Katty!" said he, "they're too young to know one school from another—at least a bad one from a good one. Just let us ask them quietly how they came to go there, instead of comin' back with their books, for as Nanny says, they never went in of themselves, as the place was strange to them. Let us all go in first."

Having got into the house, Phil seated himself with a most magisterial air, and the others stood around in anxious expectation, Katty especially. "Come here now, Jemmy!" he said to the eldest, "and tell me who it was that asked you to go into that school-house."

"It was the big lady, sir," answered the boy, still sobbing with fear.

"That's Mistress Perkins," said Phil—nodding to the women. "Well! and where did you meet her?"

"Jist at the cross-roads, sir, as we were turnin' down to our own house."

"An' what did she say to you?—don't be afraid, now—speak up like a man."

"She—she—she asked us, sir, was—was our

4*

father livin', an' we said no, he wasn't—an'—an
then she asked us about where mammy was, an' if
we weren't hungry, an' when we said no! we
weren't hungry *now*, for we got our breakfast *this*
moruin', she asked us where we were goin', an'
when I said that we were goin' nome for our books
to go to school, she said we were very good boys,
that she liked boys that went to school, an' she
said to never mind goin' home for our books, for
that she had very nice books for us—an' says she,
'I'll walk back with you myself, till I tell the master
what good boys you are.'"

"An' then she took you to Jenkinson's school—
eh, Jemmy—why didn't you tell her, that it was
to the priest's school you were goin'—

"She didn't ask us, sir, what school we were
goin' to, but when we went in, Terry here was
afeard o' the strange man that was there—him
that thought to keep us from you, sir—but the big
lady gave us both some nice sugar-stick, an' tould
us we mustn't be afeard o' the new teacher, for
that he'd be very good to us, an' show us how to
read the purty picture books that she gave us.
An' sure we thought it was the school, sir, or
we wouldn't go."

"So it was the school, Jemmy!" said Phil,
laughing, "but not the school you are to go to—
there's more schools than one, my little man! an

mind you never go near *that* school-house again.—neither you nor Terry."

"No, sir!" said both boys, " an' sure you'll not let mammy bate us—sure you'll not, sir?"

Being assured that they had nothing to fear, the children ran off, with light hearts and lighter heels, to play with the baby who was rolling about on the floor. Nanny sat with her hands clasped over her knees, and her eyes staring wide open. "Well! the Lord be praised!" she cried at last, drawing a long breath, "if them Bible-readers arn't the quarest people in the world wide. Now what good does it do them to be kidnappin' the children this way—don't they know in their hearts that the Scripture never tould them to desave innocent children, an' carry on as they're doin' in every way?"

"Faith, misthress," said Patsey, "if it does it can't be God's book, an' the priest says it is, so I'm sure they don't go by its biddin' any way"—

"For all that they have it at their fingers' ends" —chimed in Katty—"there's not a thing they do, good, bad, or indifferent, but they'll tell you, 'it's written in the Bible'—*inagh!* but it's a quare use to make o' the word of God, to be hittin' poor cratures like us with it, right an' left. But thanks be to God," she added, going back to her work, "thanks be to God that the masther took the trouble of goin' after the children, or it's what they might be goin' there for many a day without me knowin' what school they were at.'

"How did the Prodestan' soup taste, Katty aroon?" said Brian, winking at Phil. "They tell me it's very nourishin'."

"Behave yourself, now, Brian!" said Katty, and her pale face was instantly covered with blushes. "If you want to know how it tastes you can go yourself an' get some."

"An' how do you know but it's what I'm thinkin' of doin'? I only want to know does it take much of it to fill one's stomach"—casting a comical glance at his own—"mine houlds a good deal, an' I'm afeard they'd be expectin' too much from me, if I wanted it filled. I suppose now, Katty, the more soup an' bread a fellow wants, he has to go the farther agin Popery—as they call it. Now what do you think a tall, raw-boned lad like me might have to do, in case the hunger drove me over to the soup-house some fine mornin'?"

"God grant that you may never be brought to that, Brian, nor any one that I wish well!" said Katty, in a tone of deep feeling. "The Lord he knows their soup's as bitter as gall to the most o' them that takes it, for every mouthful goes agin their conscience."

"Go off to your work now, both of you!" cried Nanny, in he bustling way, "an' don't be makin' your game of poor Katty for what wasn't her fault!"

"True for you, misthress!" said Patsey, as they both took up their caubeens—"it's no laughin' matter, afther all, for it's either to stay at home an' die of hunger, or go to the Jumpers and sell one's sowl—may the good God save everybody's rearin' from such a hard fate as that!—come away, Brian—we're losin' our time, an' we oughtn't to take advantage of the master's goodness to play on him!—come away!"

"Here I am, my honey! jist alongside o' you—bedad it's the greatest pity in the world you weren't a Jumper yourself—my sowl to goodness, Patsey! but you have a great tongue in your head—why, man! you'd be worth a mint o' money to the Bible-men—agh! if I could only preach such a sarmon, it's not handlin' the spade I'd be any way!" So saying he vaulted out through the door-way, making a grimace at Patsey behind his back that set the others a laughing heartily.

CHAPTER III.

*" Thy gates open wide to the poor and the stranger—
There smiles hospitality hearty and free "*

THERE was heavy sorrow in the house of Ber
nard O'Daly when Cormac announced his inten
tion of going to America. The girls thought, at
first, that he spoke only in jest, and the old woman, from her high-backed straw chair in the
chimney-corner, loudly declared her incredulity.

" I'll never believe it, Cormac," cried his mother, " I'll never believe that you'd go away an'
leave us in our hour of need, you that's the greatest support we have; oh, no, Cormac, you'll not
do *that*, any way !"

But when Cormac went over and sat down beside her, and took her two thin, wasted hands in
his, and squeezed them hard, without uttering
a word, then the poor mother understood the
mute eloquence of her son's eyes, and she burst
into a passionate flood of tears, and for some time
refused all consolation. In vain did her daughters
and her husband try to comfort her—she would
only put them away with her hand, and cry all
the more.

" Mother," said Cormac, his own eyes testifying
how deeply he sympathized with her sorrow—
" Mother, dear, you'll ruin yourself if you go on
in that way ; you know very well that I wouldn't
leave you on any account, if it were not that I
hoped to benefit your condition by going.'

" Agra gal !" sobbed his mother, " you'll never
benefit *mine* that way, whatever you may do to the
others ; I'll not be long a trouble to you or them.
The heart within me is dead and cold, an' if you'll
only wait a little, Cormac, you'll see your poor
old mother laid decently in the grave, an' then you
can go—but don't go till then, my son--don't,
alanna machree! don't, an' God bless you !"

Cormac knew not what to say ; his sisters were
weeping around him, and little Eveleen had her
arms twined lovingly around his neck, in wordless
entreaty—he looked at his brothers, but their own
hearts were heavy, and they had not a word to
say. " Father !" said the agitated son, " will you
not try to soothe and comfort my dear mother—
oh! father, dear, won't you tell her how we must
all starve if something be not done, and that
speedily ?"

" I will, my son—I'll do my best !" and the
poor old man wiped his eyes, and tried as well as
he could to curb his feelings. " Now, Honora,"
said he, laying his hand tenderly on hers *where*

they lay clasped on her knee, " Now, Honora, lis ten to me !"

" I am listenin'," said she, without looking up.

" Don't you know, *astore machree!* that we're going to be put out of our place—the old place that my father an' my grandfather had before me —well, what are we to do then, Honora, dear! if we havn't something to look to for support? Jist think a minute, now, an' call to mind how many of our neighbors that were well off a few years ago, are now either in the poor-house or beggin' their bit, an' some of them"—he involuntarily lowered his voice—" some of them died of want an' hun- ger—think of that, Honora! your time or mine won't be long either one way or the other, *aroon!* but what's to become of the children, Honora— them that we reared so tenderly, and were—ay, an' are so proud of?"

" Och, Barney, Barney!" sobbed the heart- broken mother, clasping her hands wildly above her head—" isn't it the black, black picture you're puttin' before me?"

" But still it's the truth, Onny, dear ; now, you see, if Cormac could only get away to America — he's young, an' strong, an' active—an' has the larnin', too, thanks be to God! an' in a year, or at most two years, he'll be able to send us what will rise us out o' poverty—until then some of the boys an' girls can go out to service—"

This was worse and worse, for poor Honora had
no small share of family pride. "What's that you
say, Barney O'Daly?" she said quickly, dashing
away her tears; "sure there never was one of
our people on either sides, at sarvice since the
memory of man. Tut, tut, man—you know that
well enough, an' where's the use of talkin' that
way—och wirra! sure I'd joyfully let them work
the nails off their fingers at home with ourselves,
sooner than that they'd be on the stranger's floor,
or any one have it to say to them—" Another
burst of crying followed, and then Bernard renewed his attack; after a little, Honora became
somewhat calmer, and then she suddenly asked her
husband was there no way of keeping Cormac.

"None in the world, Onny, except we could get
Mr. Ousely to do something for us—and that's
next to impossible."

"It's totally impossible, father," said Cormac,
"for I tell you now candidly, that I have gone to
him time after time, unknown to you, and even
got Mrs. Ousely to reason cases with him, but all
was of no use—you might as well think to drain
Lough Corrib with a spoon, as to soften his heart."

"I know that, my son—God help me, I know it
well!" said his father, with a sigh; "many and
many's the time, since I came to poverty, I have
stood with my hat in my hand among the poor
cratures at the gate, waitin' to get a word with

him as he'd be goin' out or comin' in, an' he'd
snap at me as if I was a dog, an' bid me get out of
his way, an' not be annoyin' him, unless I had
some money for him. Ochone! ochone! an'
you'd see the poor people about me how they'd
grieve to see me trated that way, but sure they
were all as badly off as myself, an' dar'n't speak
a word for me."

A fresh burst of weeping followed the old man's
words, for his children were all shocked to hear
how their beloved father had been obliged to humble himself, and for their sakes, and all were thinking of the time, but a few short years before, when
Mr. O'Daly was a man of great influence with the
landlord, and was wont to be ushered into the parlor at Ousely Hall, when others of the tenants
were left outside. Alas! for the grievous change.
The young men in particular were heart-struck,
and their indignation knew no bounds. "Ah!"
muttered Daniel, between his teeth, "will there
ever come a time when these heartless tyrants shall
be humbled?"

"There will—be assured there will," said Cormac; "God would not be a just God—blessed be
his holy name—if the haughty, the relentless landlords of Ireland were not scourged, and with a rod
of iron. Their time will come, Daniel—never fear!"

"Why, mother, dear, you've got very quiet all
of a sudden," observed Eveleen, with the playful

privilege of a petted child, "are yo. going to let Cormac go ?"

All eyes were now turned on the old woman, and sure enough there was a wondrous change in her manner and appearance. Not a tear was in her eye, and it was only the increased paleness on her furrowed check that told of the recent storm of feeling. "No, Eveleen," said she, patting the little girl's head—it was a beautiful head, too, with its long fair tresses, "No, Eveleen, I'll not give my consent till I take some time to think the matter over, an' pray to God to direct us all for the best. An', children, I put it on you all to pray this night with that intention. We'll say no more about it now."

And no more *was* then said on the subject, but there was something in the old woman's manner that excited the attention of the whole family, and very often during the evening the young people talked it over. "I'll lay a wager," said Daniel to his younger brother, Owen, as they strolled out together through the green fields and down by the banks of the gurgling rivulet, "I'll lay a wager that mother has something in her head, for if she hadn't, she'd never get so calm all in a minute. I wish we could know what she's up to."

"Well, I'd give a trifle to know myself," said Owen, who was a fine, well-grown lad of sixteen, "but there's no great use in puzzling our brains about it—who's that coming up the road there?—Why, I declare, Dan, it's old granny Mulligan—hurra! let us go to meet her."

Off ran the two lads, bounding across ditches and hedges like young antelopes, till, jumping the last fence, they alighted on the high way, right in front of the individual in question, who was a little old woman not much more than four feet high, with a keen shrewd eye, and a rather intelligent cast of countenance. She was clad in an old red cloak, and a dark-colored gown of that home-made stuff known amongst the Irish peasantry as *drugget*. On her shoulders she carried a large bag, while a smaller one hung from her apron-string—she had an apron of coarse blue linen. Her feet were cased in good strong shoes, and she stumped along supported by a stout oaken cudgel. There was altogether a look of cleanliness and of self-respect about the old woman, with a sort of masculine independence in her air and bearing. Granny Mulligan was the type of a class now fast disappearing—I might almost say, gone from amongst the Irish people—she was a beggar-woman of old standing and high consideration in the district over which her rambles extended.

"Hillo! granny," cried Daniel, as he reached her side, panting and breathless; "so you've got back again. Why, we were beginning to be afraid that you'd come me no more."

" Well, you see I did, Daniel—and Owen, too—musha! give me the fist, boys; an' how's every inch of you—an' how's all at home?"

" All well—only mother's just the same way—but then she's no worse."

" God be praised, dear, God be praised!"

" Take care, granny, take care!" cried Owen, laughing as he spoke, " it's not the fashion now tc speak that way."

" What way, *agrah!*" inquired the beggar-woman.

" Why, to be praising or thanking God, or the like—if the Jumpers hear you at it they'll call you all sorts of hard names."

" Oh! the curse o' the crows on them for Jumpers!" cried granny Mulligan; " I'm blest an' happy, boys, but my heart's broke with them."

" Why, how is that, granny?" and one winked at the other, having heard the old woman's grievance at least a score of times—" What's wrong now?"

" What's wrong, is it? There, Owen, I see you want to relieve poor granny—God mark you with grace, child; many's the time you carried the bag for me before now!—well! Dan, *avick!* you asked me what was wrong now, an' I tell you all's wrong with us poor travelin' cratures. There never was luck or grace in the counthry since them Bible-readers got their grip on it. People will be

talkin' about the famine, an' the famine, but I
tell you this, Daniel O'Daly. them black-faced
fellows with their smooth tongues an' their bundle
o' books undher their oxther, an' the whites o' their
eyes turned up like a duck in thundher—it's them
hat's the rale curse o' the counthry! ay, indeed!
worse than the famine fifty times over."

  " Tut, tut, granny, you don't say so— why, what
harm do they do you or the like of you, so long as
they don't get you to turn?"

  " What harm, *inagh!* why, they do us this
harm," said granny, warmly, "that they close the
hearts o' the people agin us, tellin' them that it's
in the poor-house we ought to be, an' that it's not
good to be encouragin' us in idleness—an' that
we're a burthen on the counthry, an' all sich
things—oh, then! oh, then!—God grant me patience—was there ever sich times in Ireland as the
good ould times when there was neither poor-houses
nor Jumpers, nor Bible-readers—an' when the
poor travelin' cratures had a welcome in every
house, an' a seat at every fire-side, an' the best bit
an' sup that was goin'!—ochone! ochone! there
was no sich thing as famine or starvation in them
days—an' what's more, there wouldn't be any now
if it wasn't for the poor-houses, an' the Jumpers—
the hard-hearted haythens, that's puttin' the ould
warm charity out of the people's hearts, an' bringin' down the black curse on the counthry!"

" Well, 1 do believe that you're saying the truth, granny !" said Daniel. " Jesting aside, there seems to be a curse resting on the country ever since these scheming vagabonds settled in it —but here we are, just at the house, granny. I've a favor to ask of you, before we go in."

" Ah, then, what is it, *ma bouchal bawn ?*"

" Cormac is trying to get my mother's consent to go to America, and my father and the priest thinks that both he and I ought to go, but my poor mother doesn't know yet that there's any one but Cormac in the notion of it—now, mind, you must put in a good word for both of us."

" Well, it's like I will, Dan, agrah; for when Father O'Driscoll an' your father has it made up atween them, it must be for the best, an' we *must* get your mother brought roun' one way or an-other—though, God pity her, it'll go hard with her—but then, what must be *must* be ! Husht, now, boys," there was no one speaking but her-self—" husht ! not a word now !" so in she marched with the step of one who felt herself at home.

" God save all here !" said granny.

" God save you kindly, honest woman !" replied Bernard, who was smoking his pipe in the corner.

" 'Deed, an' you used to know me better than that, Bernard, said the old woman, throwing back her hood.

"Why, bless my soul, granny Mulligan, is this you?" cried Bernard, coming forward with outstretched hand. "Honora! Kathleen! Bridget! where are you all gone to—sure here's granny Mulligan!" Out ran the girls from an inner room, and their mother was not long behind. Eveleen caught the old woman round the neck, and kissed her over and over, saying—"Granny, dear! what in the world kept you so long away from us—why, I didn't hear a story this ever so long, for nobody tells me any when you are away!"

Before the greetings were all exchanged, Owen and Daniel came in, the former setting down the bag in a corner with a great swing. "An' why don't you welcome me?" said he with a merry laugh, "sure it's me that carries the bag, don't you see, so granny an' myself's in partnership!"

"Get out, you young scape-grace!" said his mother, "who'd be for throwin' away a welcome on the likes o' you!" and her dim eye was for a moment brightened, as it rested with maternal pride on the handsome, roguish countenance of the light-hearted boy.

"Come an' sit down here beside me, *aroon!*" said Honora, "till we have a little *shanachus.*"

"'Deed an' I will, Mrs. O'Daly, an' glad to sit down too, for I ve walked a good six miles since mornin'. Here, girls, I see you're waitin' for my duds—stop, Bridget! *aroon*, I'll give them to

Eveleen—there now, Eveleen dear, put away granny's red cloak—an' there's my meal-bag, Kathleen—hang it up there in its ould-place beside the salt-box!" When all was done as she desired, and granny comfortably settled beside Mrs. O' Daly, with Eveleen on a little creepy at her side, there were a thousand questions asked and answered, and many an exclamation of pity and of wonder escaped the listeners, as the old woman detailed how this family had been put out of their land, and were living under a shed by the way-side, and how that other had to go to the poor-house— how this one had died of starvation, and the other was "lyin' in the fever." But ever and anon granny Mulligan's eyes wandered over the kitchen and its "plenishing," and in the midst of her narration, a sigh would come, for there was indeed "a change on all things." In vain did she look for the flitches of bacon and goodly hams, and smoked heads, which in other days hung suspended from the smoke-blackened rafters—the nicely-cleaned milk-vessels were ranged along under "the dresser," but granny was sorry to see so many of them, for it showed that there was now no other use for them —the cows that used to fill them were all gone— sold to make up the rent, and all in vain—the rent wasn't paid, nor couldn't be paid, as granny Mulligan well knew.

"But where's Cormac?" said the beggar-woman

I'm thinkin' long to see him;—poor fellow! they tell me he's in a notion of goin' to America."

The words were scarcely uttered, when Cormac himself lifted the latch and walked in, his face flushed, and his eyes sparkling, like one who had been recently engaged in some angry contest.

"Speak of the devil and he'll appear," said Kathleen. "I believe if you'd spoken sooner, granny, Cormac would have come sooner. See who's here, Cormac!"

The young man no sooner saw granny than the angry frown was gone, and his face lit up with a cheerful smile. Going over to her, he took hold of her proffered hand and shook it warmly. "You're welcome back, granny," said he, "and I'm sorry we have not as good a way for you as we used to have—times are changed with us, granny Mulligan! even since you were here half a year ago. Still I'm glad to see you, granny—indeed I am. Where did you leave your daughter Aileen, or how is she ?"

"She's well, I hope," said the old woman with a sudden change of countenance; "I trust in God she's well, for she's gone—gone, Cormac; that cough that she had so long, turned into a decline, an' she's lyin' below in Tullyallen church-yard—"

"What? you don't mean to say that she's dead?" cried Cormac and Kathleen in a breath.

"Ay indeed do I, children!" said granny, the

big tears coursing down her wrinkled cheek—" she
died three months ago, an' I had hard work to get
a coffin for her—only for Father Dempsey, the
priest that's there, I couldn't a managed it, but he
got a coffin himself, may the Lord pour down
blessin's upon him, an' so I put my fair-haired
colleen into it, an' a dacent man that's there—Paddy
O'Carolan by name—put it across his horse's back,
an' him an' his son an' myself went with it,—an'
he dug the grave himself for me—an' between us
three we lifted poor Aileen into the grave, an' poor
granny Mulligan was left all alone, without friend
or fellow in the whole wide world !"   Putting her
blue apron up to her eyes, she wept for some time
unrestrainedly, for all felt that her grief was sacred,
but when she began to wipe her eyes and clear her
voice, then every one offered some kindly word of
comfort, and the old woman, by a strong and
characteristic effort, drove her grief back into her
desolate heart, and asked Cormac what was the
matter with him when he came in.

"Ay, indeed, Cormac," said his father, "there
must have been something wrong, for you looked
wild and quare somehow."   His mother opened
her eyes wide, and looked intently at her son, and
his brothers and sisters crowded around in eager
expectation.

"It was on_y a trifle after all," said Cormac, with a
smile, "so you needn't look at me as though I had

seven heads on me. I was just turning out of
Phil Maguire s lane, when who should come up but
Andrew McGilligan, the Bible-reader, with a bundle
of tracts under his arm. I nodded and bade him
'good evening:' and was passing on, but well becomes Andrew, he pulls out a tract and offers it to
me. I asked him what it was, pretending I didn't
know. ' It is a mouthful of food for the famishing,'
said Andrew, ' take and eat, and be filled.'
'Thank you very much,' said I, 'but I really am
not one of the famishing—so you must excuse me !"
and again I would have passed him, but he was not
to be so easily shook off. 'Young man !' said he,
in a very solemn voice, 'you are not sensible of
your wants, and they are, therefore, the more
grievous. Take what I offer you—read—and you
will then see how blind and ignorant you are.'
'You are certainly very polite,' said I, ' to say the
least of it, and you are also very presumptuous,
my good sir, to suppose that *you* can enlighten me
—as for your tract there, I might, to oblige you,
take it home to light my father's pipe, or even
dispose of it more quickly, by tearing it in pieces
and flinging it to the winds, did I not know that
every tract you get rid of is a victory gained.
You will oblige me by taking your way in peace
as I shall take mine. I want no conversation with
you.' ' You are very uncivil,' quoth Andrew, 'yea,
young man, you are puffed up with the pride and

uncharitableness of your religion—oh!' and A drew
groaned piteously, ' oh! what a hideous spirit doth
abide in those who follow the great delusion'—' I'll
just tell you what it is, my good fellow,' said I,
breaking in rather suddenly on his fine soliloquy,
' if you don't hold your peace, or otherwise keep a
civil tongue in your head, I'll send you headlong
into the drain—how dare you speak in that way
to one who knows both yourself and your sham
religion so well as I do?' 'Even so,' said Andrew,
moving a step or two away, ' even so were the
apostles of old persecuted—ay, verily, and the
prophets—oh! Rome! Rome! thou that dost persecute and kill—'

" ' Frogs and grasshoppers !' cried Brian Hanratty coming up behind, and giving the poor
Bible-reader such a dig in the ribs with the point
of his stick, that he roared out 'Murder! murder!'
—'oh! the devil murder you,' said Brian, 'it's a
thousand pities you *weren't* murdered—the counthry
'd be well rid of the whole jing-bang of you. I
wish to my soul that the ould boy who sent you
in among us, would jist come quietly some fine
night, an' take you back to himself.' For me, I
did nothing but laugh heartily, but Andrew began
to look very black at Brian. 'Oh! you bloodthirsty
villain !' said he, rubbing his side—' I'll—I'll'.—
Do you want another touch, Andrew !' said Brian,
cutting a caper with his shillelagh—'by the law

man, I'll tire you out before I lave you. Sure you were wantin' to convert this dacent boy, Cormac O'Daly—now why don't you thry your hand on me—eh, Andrew?' 'I'll leave you to yourselves, unhappy sons of perdition,' said the Bible-reader, preparing to cross a ditch into the fields. 'Won't you lave us a lock of your hair, Andy dear,' cried Brian, 'jist to poison the rats?— or a tract.' But Andrew was in too great a hurry to get his lank carcase out of the way of danger, so he merely turned his vinegar face, and looked daggers at myself and Brian—the latter laughed and made a grab at the bundle of tracts—the Bible-reader, who was then climbing the ditch— instinctively let go his hold, for the purpose of protecting the tracts—when his foot slipped, and down he came souse into the water, where he lay sprawling on the broad of his back, and roaring like an elephant. By this time there were several persons collected, and the unfortunate Scripture- reader was calling on one and another to help him out, but no one was in any great hurry, for they all enjoyed the fun. 'Can't you read us a chapter, Andy honey?' said one—'Won't you give me a tract, dear?' says another—'You must wait till he dries them, then,' says another; 'don't you see they're swimmin' there along side of him'—'Come, come, boys!' says Brian, 'let us take him out any way—divil an' all as he is, we can't lave him in too long We then pulled the shivering wretch out, and set him on his feet, Brian asking him very politely how he felt after his cold bath. 'Villain!' said the crest-fallen champion of Bible religion, as he shook his dripping garments, and looked ruefully down at his scattered tracts, now floating away on the stream, ' villain !' shaking his fist at Brian, ' I'll make you rue this.' ' Go to the d—l an' shake yourself, my fine fellow !' said Brian very coolly ; ' wasn't it your own fault from beginnin' to end— what business had you forcin' your bit of a tract an' your hypocritical discoorse on them that could tache you and your betthers ? Be off with you now, an' chaw your cud on the lesson you've got— maybe it'll be of some sarvice to you !' With that the Bible-reader turned off into Billy Wallace's meadow, and made for the house, while the boys stood on the road and cheered him till he got in out of their sight. So, after we had enjoyed a good laugh at Andrew's expense, I bid them all good night, and came off home, little thinking that I'd find granny Mulligan here before me."

Young and old were much amused by Cormac's account of McGilligan's discomfiture, and one and all declared that it was "good for him." Eveleen alone demurred—" not but I'd be glad to see him getting the worst of it," said she, "for many a day he teased Owen and myself to take tracts or Testaments from him when we'd be going to school, ay! and call us bad names when we wouldn't take them, but then I'd be sorry to see any one falling into the water that way—oh dear!" and Eveleen shivered as though she felt the cold in her own proper person.

"That is just like you, my gentle Eveleen!" said Cormac, as he drew the little girl to his side. "But you must remember, child! that it wasn't my fault nor Brian's neither—he merely missed his foot—trying to save his precious burden from a ducking, he got one himself."

"Sorra mend him!" said granny. "If he had been ducked on purpose it's what he'd desarve. I'm just thinkin' about a thing that happened down at Tullyallen while I was there, an' as I know you're all fond of stories, especially Eveleen here, I'll just tell it to you to pass the time."

Eveleen clapped her hands and cried out, "Oh a story! a story! dear, good granny, do tell us a story!"

"Better have supper first, Eveleen!" observed Kathleen; "move round there, boys, till Bridget puts in the table—here's the porridge ready. Granny will be in a better way of telling the story when she has had something to eat and drink."

"That you may never have worse news for us, Kauth!" cried Daniel, as he and Owen pushed back their chairs to make way for the table.

"We havn't many dainties to offer," observed

Honora, as her daughters dished the homely meal —" but, sit over, granny! an' take some supper!" A large bowl of milk was placed before the guest, but granny's keen eye soon saw that the liquid in the tin 'porringers' of the others was not all milk, being diluted with water to make it go the farther. Bridget noticed the look with which granny followed her motions as she prepared the beverage, and a smile dimpled her rosy cheek as she remarked: " We havn't so many cows, granny, as we used to have—the *seven* are reduced to *one*." " Well! God's will be done, Bridget!" said the old woman with a heavy sigh. "More was the pity that your store 'ud ever be less! But never mind, *agrah!* never mind—there's a good time comin'."

Eveleen kept watching the progress of the meal with great impatience, herself was the first to push back her seat, and when the others had nearly all followed her example, she was somewhat indignant to see that Owen and Daniel were still masticating. "Why, then, I'm sure you might be done, now," she said to them, "for I do believe you were first at the table. Can't you swallow down quickly, till we get the table away—now if you don't make haste, we'll not have the story to-night, for granny will want to go to bed soon."

" Here, then, girls," said Daniel, the last to rise, " come along and take away the table—poor

Eveleen must have the story." So the table was
removed, a fresh fire made, and the hearth swept
nicely up with the heather broom that stood in
the corner, then the whole family gathered around
—Eveleen taking her usual station at granny's
nee, and the old woman began her narrative.

"About two months agone," said she, " there
was one o' the paupers—as they call them, with
their new-fashioned names—took sick in the poor
house below at Tullyallen, an' she got so bad all
of a suddent, that the nurse sent off for the priest.
Well! you see, the poor crature couldn't spake a
word, an' one o' the officers of the house—to be
sure! took it upon himself to send off another
messenger for the ministher, bekase he said that the
woman was entered on the books as a Prodestan'.
Well! sure enough but the ministher got in first,
an' he was just a goin' to kneel down an' pray—
sure that's all the man could do—when the door
opens, an' who walks in but the priest, as tall an'
as straight as a may-pole, my jewel! So he went
over, an' took hould o' the woman's hand jist to feel
her pulse, before he'd do anything, an' up starts
the ministher to his feet: ' what brings *you* here?'
says he, quite sharp and crusty. ' My business!'
says his reverence; ' what brings *you* here?'
I was sent for, sir,' says the ministher. ' And so
was I,' says the priest back again to him. ' Isn't
this woman a Protestant, my good girl?' says the

ministher, turnin' round to the nurse that was in it.
'I don't think she is, sir,' says the nurse, 'for I
got a pair o' beads in her pocket.' 'Well! at any
rate,' says the ministher, says he, 'I was sent for,
an' I'll do my duty.' 'An that's not much,' says
the priest, with a kind of a smile; 'but the best
way to settle the dispute is to ask the woman
herself—perhaps she can speak that much.' So he
stoops down to ask the sick woman if she wasn't a
Catholic, an' well becomes the ministher, didn't he
give him a pounce right on the back o' the neck
that bobbed his head down on the woman's breast."

"Oh, the villain!" cried Bernard. "The bad,
bad man!" said Eveleen, "but what did the priest
do then?"

"What did he do?" said granny, with a smile,
"why, he jist got up, an' turn'd on the ministher,
an' gives him one box of his big fist that sent him
spinnin' like a top across the room."

"My hundred blessings on him!" said Honora;
"that was just the way to sarve the villain—"

"Yes," observed Cormac, "for argument is
thrown away on a lad like that—but what followed,
granny? I hope the priest kept his ground beside
the sick bed."

"Indeed then he did, Cormac, an' he took the
ministher coolly an' quietly an' put him outside
the door, when he was goin' on talkin' an' makin'
a noise—ther his reverence gave the rites o' the

church to the poor woman, an' went his way home.
Well, what would you have of it, my dears! but
the ministher summonsed the priest for an assault,
an', bedad, when the priest seen that, he thought
he wouldn't let it all go for nothing—an' didn't he
summons the other ; an' sure enough it was the
ministher gave the first assault.   Well, bedad, the
day came, an' away goes them all to the coort-
house, an' there was a good many brought there
for evidence, but amongst the rest was the nurse,
a fine, stout, rattlin' girl as you'd see in a day's
walkin'.   Well, she went up on the table, to be
sure, to give her evidence, an' who should be
standin' beside her but the crier of the court, a
little, weeny bit of a man, with a lame leg, an' an
old withered face on him that wasn't a bit bigger
than the palm o' my hand.   The magistrate began
to put questions to the girl, an', of coorse, she answered them ; an' at last, they ax'd her how did
the ministher hit the priest.   ' Why,' says she, ' he
jist took him by the back o' the neck, this way,
your honor,' an' she catches the little man along
side of her by the collar o' the coat, ' an' he gives
him a push this-a-way, your honor,' an' she gives
the poor crier sich another drive that down he
went headforemost among the people outside, an
with that there was sich a shout of a laugh all over
the coort-house that you'd hear it a mile off, an'

indeed, they say there wasn't one in it, magisthrate or else, that you couldn't tie with a sthraw."

Granny could scarcely get her story finished, with the roars of laughter that it drew forth Honora herself had to press her hands on her sides, and beg of granny to leave off, for she couldn't stand it any longer.

"It's all done now, dear," said the beggar-woman, with imperturbable gravity—"that's the whole of it."

"But, granny," cried Daniel, as soon as he could speak from laughing, "do you think did she intend to knock down the little man?"

"No more than you did, *ma bouchal!* that never seen him. No, no, she wasn't mindin' what she was doin' at all, but jist catch'd a hould of him as he was near her, to show the magisthrates how it happened, and when she gave him the shake, you see, her arm was so sthrong, an' him so weak an' donsy, that he couldn't keep his feet. Oh, hedad, she didn't mane it at all, for she was sorry enough when she found the little man gone—but, you know, it couldn't be helped then."

"Well, really, that's a good story," said Cormac, "the best I've heard for many a day—what do you think, Eveleen?"

"It's very funny," said Eveleen, "and I couldn't help laughing at it, but I hope the poor little fellow wasn't hurt—eh, granny?"

"Oh, not much, I b'lieve—only a good deal frightened. But, now, I think it's bed-time, and I was up at day-light this mornin." This was the signal for a general move; so the night-prayers were said and all went to seek repose.

CHAPTER IV.

When man has shut the door making
On pity, earth's divinest guest,
The wand'rer never fails to find
A sweet abode in woman's breast—CARCANET.

THE dawn was just beginning to shed its crimson
light over the eastern hills, and the earth was still
silent, when the door of Bernard O'Daly's house
was softly opened, and two female figures issued
forth, carefully wrapped up in large gray cloaks.
One was old, or at least infirm, for she leaned
heavily on the arm of her companion, whose light
step and slender proportions were those of the
spring-time of life, but the faces of both were par
tially concealed by the hoods of their cloaks.
We may as well anticipate our reader's suspicions,
and announce that these were Honora O'Daly and
her daughter Bridget. But why were they abroad
so early, and evidently unknown to the other
inmates of the house ? Let us follow them on
their way and we shall see. Scarcely a word was
said by either, as they followed the upward course
of the rivulet for about a mile, and then turned off
through the fields till they came out on the high-road
in front of a handsome gateway of cut stone, with

a small but beautiful lodge at one side within, and a smooth, well-kept avenue, with its fringes of green, winding far and away between rows of tall sycamores, intermingled with beach and ash. Long did our two lone wayfarers wait outside the gate before any one was stirring, but at length the door of the gate-house opened, and a tall, lazy-looking fellow made his appearance, stretching and yawning as though he had not slept enough. He was moving away in another direction, around the end of the lodge, when Bridget called to him through the gate, "Larry, I say, Larry!"

"Who's that callin' me?" said Larry, coming towards the gate.

"It's me, Larry—Bridget O'Daly. My mother's here too, so make haste and let us in."

"Oh indeed an' I will then," said Larry, as he leisurely took down the huge key from a nail inside the lodge-door, and proceeded to open the gate. "But what in the world wide brought ye out so early this mornin'?—why myself 'ud sleep this hour if it wasn't that I'd be afeard o' the masther comin' down!"

"Larry!" said Honora, speaking now for the first time, "I want to see Mr. Ousely, an' as I know he's a very early man, I thought I'd come before there would be anybody else here, or that he'd be goin' out some place for the day."

"Well! I don't know," said Larry musingly

"he'll not be plased at me for lettin' any one in so early about business—of coorse, it's on business ye're comin'. Mrs. O'Daly?" and he looked searchingly under the hood, for it still shaded her face.

"Why—yes—it's on a little business of my own that I wanted to see him."

"Humph!" says Larry, putting his finger to his nose, "Humph! I know—well! now, Mrs. O' Daly, I'll just ax you one question before you go to thry your fortune—will you be willin' to pass yourself off as if you were thinkin' of turnin'— tell me that now?"

"Larry Colgan! I hope you don't mane to insult me—I didn't expect it from you."

"No offence, ma'am, no offence—I'd be long sorry to offend you—but answer me the question I put to you."

"Well! if I must answer such a foolish ques- tion, I say 'No!' not for all Mr. Ousely's worth!"

"Very good—that's just what I thought—well, then, ma'am, you may as well turn straight back, for you'll only make matters worse if you go. Take a friend's advice an' go home."

"Why, Larry, you're makin' the devil blacker than he is—sure if the master was so inveterate as that against us, he wouldn't have *you* here."

"An' if he has me here," retorted Larry with a chuckling laugh, "it's bekase he thinks that I'm

7

goin' to turn—he has me on the hook, ma'am, fot
the last two years—ever since he got so black
agin Catholics, by manes of the Bible-readers—
bad manners to them—'deed he has, ma'am, but
somehow he never gets me a-shore, for I'm able
enough for him one way or another. The only
thing is that I don't get goin' to chapel, but then
when I don't go anywhere else, I have hopes that
Him above won't be hard on a poor fellow that
has a wife an' five little ones to keep the life in."

Honora shook her head. "Take care of that,
Larry, I'm afeard that such excuses as them won't
save us—but, after all you tell me, I'll venture up
when I've come so far—an' och! och! but it's the
heavy thrial that's before me. Is the mistress
likely to be seen at this hour ?"

"Or Miss Eleanor ?" said Bridget.

"Well! I don't know that you'll see either of
them—though you might, perhaps—for Miss
Eleanor is an early riser, an' God bless her every
day she *does* rise—there 'ud be no livin' here of
late days only for her."

"Good bye, then, Larry. good bye," said
Bridget, " till we see you again."

"Mind what I tould you, now!" said Larry,
calling after them. "You'll say I'm a prophet, I'm
thinkin', before you're either of you much oulder."

As the mother and daughter wound their way
along the nicely-sanded walk, they discoursed in

ow whispers, looking cautiously around to see that no one heard them. As they approached the house, Honora's heart sank lower and lower, and it required all Bridget's strength to support her. "Och! Bridget! Bridget!" said she, as the fine old mansion stood full before them, its numerous windows reflecting the rays of the rising sun, " havn't they heaven on earth that live in such houses as that, with such a place as this all round about them !" and she cast her heavy eyes around on the grand old oaks, and the soft verdure of the sloping lawn, and the rustic seats placed here and there under the shade of spreading branches. " Not that I'd covet to live in such a grand place," she added, " but only that I'd be able to keep my children all about me, an' to make my soul* in peace."

" And still they have their own troubles, mother —these grand quality—just as well as we have— but you see there's no one a-stir yet—will you sit down, mother dear, on one of those seats—I know you're not able for such a walk as this !"

" I know that, dear, but you see how God gives strength to the poor, weak creature in the time of necessity. We'll jist sit down on the steps here, an' then we'll be sure not to miss the masther when he comes out."

They had barely waited a few minutes, when the

* Work out my salvation.

door behind them was thrown open, and a great
black poin.er darted out, gambolling and frisk-
ing over the lawn. Honora and her daughter
stood up quickly, and, turning round, found
themselves face to face with the arbiter of their
fate. He was a stout, square-built man of middle
size, with large, coarse features, garnished on either
side by enormous black whiskers. His forehead
was low, and by no means what is called intellectual.
Still the expression of Mr. Ousely's face was not
bad, being characterized by a sort of jovial and
rather frank *bonhomie* that made some amends for
the fierce, bold look, and the flaming color.

"How now?" cried the lord of the manor
taking the two females before him for mendicants
—"what the d—l brings you here so early?—
can't you go round the other way if you want
help?"

"Oh! Bridget dear! hould me up, or I'll fall!"
whispered Honora O'Daly to her daughter; then
raising her voice as high as her weakness would
permit, as she saw that Mr. Ousely was for passing
out: "We're not beggars, your honor, though, God
help us! we don't know how soon we may be!"

"What the deuce are you then?" cried the im
patient landlord, turning short round—"what brings
you here? Speak quickly, woman? for I can't
s'and here waiting—what do you want?"

"Mr. Ousely!" said poor Honora, in her low.

murmuring voice, "you used to know me better nor this. I'm Bernard O'Daly's wife, your honor, an'—".

"The d—l you are!" cried Ousely, "an' pray, madam, what brought you out of your bed so early? I wish you had slept an hour or two longer! what brought you here?"

Confounded by the contemptuous roughness of his manner, poor Honora could not speak, but Bridget hastily answered:

"My mother was thinkin', your honor, that if she'd come up herself and speak to you, and tell you how the matter stands, you might be pleased to give my father a little time—she thought—"

"Let her speak for herself," interrupted Ousely, "I hate second-hand stories." It was now Bridget's turn to hang her head, and blush to the very temples, and try to keep in her tears.

"I say, good woman! do you mean to keep me here all day?"

Honora cleared her throat two or three times, for she felt as though her poor weak heart were rising up, up into her mouth. "Well! I was in hopes, Mr. Ousely, that if I'd come up myself—an' God knows it's ill able I am, for I didn't set a foot outside the door these six weeks—and tell you how distressed we are, you'd maybe be good enough to lie back a little longer. If we had any prospect of bein' able to keep the farm. the boys

7*

would all stay at home an' work hard, as they al
ways did, to get the arrears paid up, an' keep us
in it, but if you're goin' to put us out, an' the agint
says you are, then poor Cormac is for goin' to
America, an' maybe Daniel too, and that would
break my heart, Mr. Ousely, indeed it would, sir!
Och, sir dear! sir dear! you didn't use to be so
hard upon the poor cratures that's thryin' all they
can to plase you, an' to pay the rent as far as
they're able!"

"This is all very fine talk, Mrs. O'Daly, but it
won't do. Money I want, and money I must have
—if your husband can't give me all the amount of
the arrears, let him give me the half of it—there
now, that is a fair proposal!"

"Ah, Mr. Ousely dear! but it's you that knows
little about how we're situated, or you wouldn't
expect money from us at this present time. You
might just as well thry, your honor, to get it out
of a whin-stone!"

"In that case we are both losing our time—I
can do nothing for you—and mark me, good
woman! your son shall have some trouble in getting
away. I have heard a bad account of his conduct
lately."

"Is it Cormac, your honor?" cried the astonished
mother. "Why, who dare say anything agin his
character—oh! your honor's only jokin' I know—
sure the whole country can tell you that there's

not the likes of him in it for sobriety an' indus-
thry, an' for a good son an' a good brother, the
Lord never put the breath o' life in a betther boy !
Oh! Mr. Ousely! say what you like to me—I can
bear anything, anything at all, but don't spake agin
Cormac—I can't stand that, your honor, for that
boy is the pride o' the whole family—"

"We shall soon see that !" said Ousely, cutting
her short, " you may go now, for you have your
answer !" He turned away, and began to whistle
for his dog.

" Well! Mr. Ousely," said Honora, in a firmer
tone than she had before spoken, "I suppose I *may*
go. I stole out this mornin' before any of our
people were stirrin', for I knew they wouldn't let
me come on such an errand. I'm goin' back to
them in bad heart, but I have this comfort, that if
there's no pity or mercy for us in this world, there
is in the next—God sees all this !"

" Oh certainly, and so does the Virgin Mary—
here, Prince, Prince ! Now, my good woman, it's
a pity it wasn't to the Virgin you applied in this
emergency, they say she's great at working miracles
for you Papists !"

Shocked by the contemptuous tone in which he
spoke of the Blessed Mother of God, the poor
woman was moving away without any reply, but
a bright idea had en/ered Ousely's sluggish mind.

and he was now intent on carrying it out, so he was at her side in an instant.

"I say, Mrs. O'Daly"—Honora stopped still—"what would you think, now, of coming over to us, the whole of you, and if you do—"

"I don't very well understand you, sir, I'm only a plain, simple woman, an' not used to fine Eng lish—"

"What the deuce! havn't I spoken plain enough, knowing your ignorance! I say you can get over all your trouble, if you'll only give up the old, crumbling Church of Rome, which is your ruin and the ruin of many others!"

"Oh! you're not in earnest now, Mr. Ousely, I know very well you're not!" said Honora O' Daly in a faint voice.

"Upon my honor and soul, good woman! I never was more in earnest in my life, and I speak to you as a friend!"

"Och then, the Lord deliver me from sich friends!" and poor Honora's voice sank lower and lower, till it was almost inaudible. "Come, Bridget! give me your arm and let us go, we're long enough here!"

"So you won't condescend to answer me madam!" cried Ousely, his face flaming with anger "What am I to think of such conduct?"

"Mr. Ousely!" said Honora, and throwing back her hood for the first time, she startled even

the imperious landlord by the sight of a countenance pale as death, eyes sunken and hollow, and
lips colorless as those of a corpse. "Mr. Ousely!
you may be satisfied now—you have given me
the heaviest crush of all, an' my heart's broken,
broken, broken!—och! blessed Lord!" she faintly
whispered, "but we're come low, low, low, when
they'd even offer it to *us* to turn, to sell our souls
for the bit an' sup—och *wirra! wirra! wirra!*
Take me home, Bridget honey, take me home, an'
God grant I may live to see it: I'm done now any
way!"

"Mother dear!" said Bridget in a whisper,
"won't you bid Mr. Ousely good bye? he's as
mad as can be!"

"I don't care, Bridget, I'll never spake another
word to him, if I can help it; he can only do his
worst, an' he'd do that any way. If I was dyin'
this minnit I'd lave my death on him!"

"Why what the d—l have I said to make the
old gentlewoman so angry?" shouted Ousely. "I
only wanted to put you all in the way of doing
well—upon my honor, that was all!"

"An' I'd sooner you had tramped me down in
the dust than say what you said." Honora never
turned her head as she spoke, but kept walking on
as fast as she was able.

"I tell you what, old woman," cried the angry

landlord, "you'll rue this morning's work as bitterly as ever you rued anything."

"Never!" returned Honora with an energy that
made her whole frame quiver. "I'll never rue it.
Come what will, with the help of God, I'd give
you the same answer a thousand times over. You
may put us out of the place that the O'Dalys have
had, father and son, for hundreds o' years, an' send
us to die on the road-side, or be shut up like jail-
birds in the poor-house, but that's all you can do ;
you can't take the faith from us that will comfort
us in the hour of death, an' gain heaven for us
hereafter. No, Mr. Ousely, while we have the
thrue faith, an' do what it teaches us to do, we
don't regard any one. God can bring us safe
through all, an' you can only do what He gives
you lave to do."

"Well, we shall see whether God will do anything for you or not. By my honor and word,
you'll require his aid before many hours go by.
Be off now from about the place, or I'll hunt the
dog in you!"

"Oh, mother! mother! come away," whispered
Bridget again, "he looks as if he was going to
beat us—come away fast!"

"As fast as I can, dear," said the heart-stricken
mother. "God help me! I'm a poor donsy crature! Oh, Bridget, *astore machree!* I'm afeerd

I'll never be able to walk so far, my limbs are bendin' under me."

Bridget looked round bewildered; there was not a soul in sight, for Ousely had dashed off through the trees at the rear of the mansion. "Won't you sit down on one o' these seats, mother dear, and may be when you rest a little you'd have more strength."

"No, no, Bridget, I'll not run the risk of *him* seein' me again."

"Well, then, try to keep up till we get to the gate-house, and Larry will send one of the children for some of our people to get a cart and come for us."

"I will, *agrah!*—I think—I hope—I can manage to walk that far. Och! my heart's broken, Bridget! it's down, down, never to rise again!"

"Don't say that, mother dear—oh! don't say that; I can't bear to hear you talk that way!" and poor Bridget could scarcely speak without sobbling.

Leaving Honora and her daughter to make their way home as they best can, let us return and take a peep at what was going on in the interior of Ousely Hall.

The breakfast parlor was arranged for the morning meal. A bright coal fire was burning in the polished brass grate, the table was set in front of the fireplace, and nothing could be more elegant

than the snowy damask cloth, the silver tea service
and the beautiful Dresden china. The tea-kettle
was steaming away on a stand within the fender,
and a large plate of buttered toast was placed on
a steamer close by, awaiting the time appointed
for its demolishment. The furniture of the room
was not of the newest style, but it was rich and
heavy, and well adapted to promote comfort. The
two windows were hung with crimson drapery,
which transmitted a soft warm light into the room,
that made it look still more comfortable. At first
sight there was no living creature visible, with the
exception of a small brown spaniel, a beautiful
creature, which lay on a cushion near the hearth;
but, by and by, there was a slight rustling of the
window curtains, and a young girl of some nineteen or twenty years stepped softly from behind
their folds, and threw herself into an arm chair
close by. She was a very lovely girl, with dark
radiant eyes, and a purely Grecian face, thoughtful and intelligent in expression, as such faces generally are. Her hair was of the darkest shade of
auburn, and simply braided around her finely
formed head. Her figure was slight and graceful,
and her stature considerably above the middle
size. There was a troubled and even anxious look
on her usually placid face, and she sat with her
eyes fixed on vacancy, nor moved, though her dog
went to claim the accustomed caress from her soft

nand. After a little while the door opened, and an elderly woman entered, whose face, though pale and somewhat care-worn, had so great a likeness to that of the young lady that there could be no doubt of their close kindred. They were indeed mother and daughter, the wife and daughter of Mr. Harrington Ousely.

"Why, Eleanor, my child, you look thoughtful this morning," said her mother; "what is it that engrosses your mind so much?"

"I was just thinking, mother," said Eleanor, as she placed an easy chair for her mother near the table—"I was just thinking of the very great contrast which there is between our condition and that of the poor people from whom my father draws his income."

"The contrast is certainly striking," said Mrs. Ousely as she glanced around on the luxuriously-furnished room and the elegant breakfast table; "but then it is so ordained by Providence, and we have no need to trouble ourselves about it. There have been rich and poor ever since the world began, or pretty nearly so."

"Granted, my dear mother; but it seems to me that there never were people situated exactly as these poor people are: they are starving, at least many of them are, and yet they must try to pay rent. They are patient and resigned, as none but themselves could be under such circumstances,

they never murmur against the will of God,
though it consigns them to hunger, cold, and all
manner of wretchedness,—and yet the religion
which makes them thus patient and enduring is
assailed in every shape and form. They are told
that it is superstition—folly—nay, even idolatry;
they are perishing with hunger, and tracts and
Bibles are offered them."

"Why, Eleanor," said her mother, "how can
you talk so? They are provided with good wholesome nourishment for the body as well as for the
soul."

"Yes, mother, but on what terms? Are they
not driven away like dogs from the soup-shops
unless they will consent to barter their religion, as
Esau did of old his birthright, for the mess of pot-
tage?"

"Well, well, child!" said her mother in a quer-
ulous tone, "I don't pretend to follow you in
your philosophical flights. What in the world has
started these ideas in your mind so early this
morning? You haven't been out, have you?"

"No, mother, but I happened to witness a scene
from the window just before you came in that set
me a-thinking in sober earnest. Who do you
think was sitting waiting on the steps of the hall
door this morning for my father to make his appearance?"

"Who was it, Eleanor? you know I am not very good at guessing."

"Why, poor Mrs. O'Daly."

"What, Bernard O'Daly's wife! Why, she has been ailing for several weeks. I heard it said that she was very far gone indeed."

"And so she is, mother, so far gone that I do not think she will live many days; yet you see she ventured up here, hoping that the sight of her exhausted state might move my father to do something for them. Now you and I both can understand how much that step must have cost her, for we have often been amused at the poor woman's family pride, and her efforts to keep up a show of respectability."

"Yes, I know that," said Mrs. Ousely, slowly, her features working with an indescribable emotion of pity and surprise. "Poor Honora! how low she has fallen! and indeed I am sorry—sorry. She is a very worthy woman, and brought up a fine family, I must say, although they *are* Papists. Well, did she succeed?"

"No, mother, she did *not* succeed," replied Eleanor, sadly; "my father treated her so roughly that I really felt ashamed; and as for the poor woman herself, I could see very well that her daughter Bridget, who was her only companion, had to keep her from falling. Still, if my father had contented himself with refusing her request, it would not have been so bad, but, unfortunately, he thought it a good opportunity to promote the interests of the Reformation, so he made her a proposal that if she and her family gave up their religion—paying it a very handsome compliment at the same time, such as you may well imagine—he would make all smooth, and set them on their legs again."

" Dear me, he might have known there was no use in making such an offer to *her*. I should never have thought of such a thing. Well, and how did she take it, Eleanor ?"

" Just as I expected ; her proud heart, still unsubdued, swelled up with indignation, and the effect on her feeble frame was plainly visible, even to me in here. I really think she defied my father—certain it is that she turned away, leaning on her daughter's arm, and never condescended to look at him again, though he called after her more than once in threatening language. She answered him, indeed, as well as she could, but never turned round. I would gladly have gone out to speak a word of comfort to the poor woman, but I saw my father standing looking after the two as they went down the avenue; and even when he turned away into the wood with Prince after him, I was afraid to venture, lest he might see me, for I know he would scold me most unmercifully if he saw me speaking to Mrs. O'Daly just then."

" Well," said Mrs. Ousely in a hesitating tone,

"I am just as anxious as any one to see these wretched people drawn forth from the errors of Popery, but"——

"But you could not make up your mind to outrage the unfortunate in the way that I have been describing," said Eleanor with a smile. "I know it, my dear mother, and would to heaven that my father had half your compassion for the poor; then indeed we might hope to"——

"Win the people over from the errors of Rome."

"Not exactly that, my dear mother," and Eleanor smiled again. "I meant that we might soon hope to improve the condition of our tenantry. I do not think the mire of Popery so *very* great an evil, after all. But 'tell it not in Gath,' my dear mother;" and she raised her taper finger playfully. "Hush! here comes my father. I hear him talking to Prince as though there were a dozen with him. I must hurry and put the tea to draw."

But it was not to Prince that Mr. Ousely was then talking, for he had, during his morning ramble, picked up a companion of another kind. This was a biped of the *genus* man—a tall, cadaverous-looking personage, with a singularly discontented aspect, and a pair of round shoulders that took somewhat from his unusual length and gave him the appearance of bending forward, even when he stood perfectly straight. This person was ushered into the parlor by the master of the mansion with a "Walk in, Mr. McGilligan—walk in, sir; it's only my wife and daughter."

The ladies returned the somewhat awkward bow of the visitor, and Eleanor looked inquiringly at her father. "This is Mr. McGilligan, Eleanor. Hetty, my dear," to his wife, "this is Mr. McGilligan, of whom you have often heard me speak; he is exceedingly useful to us in propagating the truth."

"I have seen the gentleman before," observed Mrs. Ousely very coolly. "Pray be seated, Mr.— Mr. McGilligan."

"I have brought him home to breakfast, Hetty," resumed her husband, "as we have some official business to transact afterwards." Politeness would not permit the ladies to express any surprise; but Eleanor could not help thinking of poor Honora O'Daly, kept standing outside the door, and dismissed with contempt and insult. She sighed as she took her place at the table and proceeded to make the tea.

McGilligan quickly perceived that the ladies were not disposed to talk with him, so he wisely addressed his conversation to his host. How great was Eleanor's surprise when she found that the excellent Scripture-reader had come for the express purpose of lodging a complaint against Cormac O'Daly and others for assault and battery. She listened with apparent indifference, but her mind was busily at work on a benevolent project

CHAPTER V.

To make one maid sincere and fair
Oh ! 'tis the utmost Heaven can do.—Meen.
Beauty alone is of but little worth,
But when the soul and body of a place
Both shine alike, then they obtain a price.—Yerme

THE only being who could really influence Harrington Ousely through his affections was his daughter Eleanor, whom he loved with nearly undivided affection, for she was his only remaining child, and such a child as could not fail to evoke all the tenderness of a parent's heart. He regarded his wife, it is true, with a sort of half-respectful, half-loving kind of feeling, and did not scold or abuse her more than once a week or so, but she had never exercised the slightest influence over him : in fact, she had not sufficient energy of character ever to make the attempt. She was naturally of a soft and yielding disposition, full of sympathy for the woes of her fellow-creatures, and ever ready to relieve them as far as lay in her power ; but she had been brought up by a popery hating old uncle and aunt, from whom she had imbibed that leading trait of character, and allowed

it to influence her whole life. Her mind had never
received any special cultivation, more than that
generally given in fashionable boarding schools, so
that her reach of thought was by no means very
extensive. Still she was a good, well-meaning
woman, and discoursed on ordinary topics with
propriety and even elegance of diction. But her
daughter was quite a different sort of person:
gifted with a high order of intellect, and a solidity
of judgment by no means common to her age and
sex, she had had the advantage of being educated
by one who was fully competent

> " To rear the tender thought;
> To teach the young idea how to shoot,
> And breathe th' enlivening spirit."

This was a widowed sister of Mr. Ousely, who had
resided in the family during the years of Eleanor's
infancy and childhood, and who, being herself a
woman of commanding talents and cultivated
mind, together with a loving and tender heart, had
elicited and matured all the higher qualities and
more amiable instincts of her niece's mind and
heart, so that when she left Ousely Hall, to take
up her abode with a sister in the south of England,
Eleanor, then seventeen, was already complete in
her education, both moral and intellectual. It was
a hard trial for both aunt and niece to tear
themselves asunder, but the path of duty must be
trod, and Mrs. Ormsby was called to watch over

the gradual decline of an only and beloved sister, who was drooping day by day, and pining away amongst strangers in a foreign land, her husband being in the employment of government, so that he could not choose his place of abode. From her earliest infancy, Eleanor had exercised no small control over her father, even when he had other children to divide his affection; but when death had gathered all the others into the dreary mansions of the tomb, then Eleanor became the reigning sovereign, and it was only when some sudden gust of passion swept away for the moment both reason and affection, that her influence failed.

On the present occasion, she made up her mind to defeat Andrew McGilligan in his vengeful machinations, but in order to do this efficaciously, it was necessary that she should abstain from any open manifestation of interest in the O'Daly family, who, as steadfast Catholics, were peculiarly obnoxious to the Jumpers. Not a word of the conversation between her father and the Bible-reader was lost on Eleanor, but she prudently refrained from joining in it, addressing herself only to her mother, and when her father, now and then, called upon her for her opinion, she answered cautiously and evasively. She was amused, however, to hear McGilligan complain of the brutal usage he had received, and suddenly raising her eyes to his face, at the close of one of his whining harangues, she asked in a cool, indifferent tone:

"It was yesterday this happened—was it not?"

"Yes, Miss, yesterday evening!" replied Andrew in his smoothest voice.

"And you were badly hurt, were you?"

"Well! no," stammered Andrew, "not to say hurt neither, but that wasn't their fault, and I was wet to the very skin."

"I do not at all doubt it," said Eleanor, drily; "a ducking involves a wetting. And so, Cormac O'Daly pushed you in, you say?"

"No, Miss Ousely, he didn't push me in, it was the other rascal called Brian—Brian something."

"O then, Cormac O'Daly had nothing to do with the ducking?"

"That's right, Eleanor!" shouted her father, "cross-examine him! Upon my honor you can do it well—keep to it, I say!"

"Oh! I have no desire to puzzle Mr. McGilligan," said Eleanor, calmly, "but it appears to me that there is no serious cause of complaint against this young man, O'Daly, and as there were so many persons present on the occasion, the truth *must* out, and the charge would, of course, fall to the ground."

"What the d—l, Eleanor!" cried Ousely, "do you mean to say that we could not give the fellow

some punishment for his impudence! a month or
two in Galway jail would cool him down some."

"Yes, but how can you have him committed to
jail? on what pretence, my dear father. If there
were any chance of the assault being proved, then
should have no objection to your receiving it, but
it strikes me that by going on with this affair you
will merely raise a laugh at Mr. McGilligan's
expense, seeing that he merely met with a rebuff
in his praiseworthy attempt to make a convert, so
far as O'Daly was concerned, and even as to this
Brian, whatever his name is, it may turn out that
even he did not mean to commit an assault. Our
worthy friend here might possibly have stept back
into the drain in the heat of the discussion."

"No, Miss," said Andrew, somewhat indignantly,
"I never forget myself so far as that; it was when
I was climbing the ditch, you see, that my foot
slipped, and even that would not have happened to
me, had it not been for that vile man, Brian, who
made an attempt to get hold of my tracts, which
being exceedingly valuable (inasmuch as there
were fifty of *the Virgin reduced to the level of Other
Women*, and seventy-five of *Confession the great
Abomination*), did incautiously let go my hold, in
my earnest anxiety to save the precious bundle,
and so fell in—

"So Brian did not actually apply hand or foot
to your person?"

"Why, he was the cause of my mishap, Miss Ousely, and if I got my best trowsers and brown surtout all covered with mud, and lost seventy-five and fifty—let me see—that is, one hundred and twenty-five of our best tracts, the fault is entirely his, aided, of course, and abetted by that incorrigible Papist, Cormac O'Daly."

"They must be made an example of, McGilligan, upon my honor and soul! they must! these stubborn Papists must be brought under, by——" and he swore an awful oath; "when neither hunger nor thirst will do it, then law *must!* that's my notion, so no more talk about it. I'll direct the clerk, after breakfast, to give you summonses for these rascals."

"You had better say nothing more about it, my dear Eleanor," said Mrs. Ousely; "the law must take its course, you know, and our excellent Scripture-readers must really be protected by the strong arm of authority, in their arduous undertaking."

"I bow to my father's superior wisdom and yours, my dear mother," said Eleanor with a smile; "and I hope Mr. McGilligan will excuse me for what I have said in pure good will."

"Oh! surely, Miss, surely," and Andrew ducked his head down on his chest, and wriggled, and smiled a wan smile. "No harm done, Miss, not the smallest!" So the breakfast went on in peace, and when it was ended, Eleanor requested her father just to look at some sketches which she was sending off to her aunt Ormsby by the first post.

" Wouldn't it do when I have got through with McGilligan ?"

" It might, but I would rather have you come now, as I am going over to Clareview this after-noon, and want to send off my letters as early as I can. Mr. McGilligan can surely wait a few minutes."

" As long as you please, miss, I'm not in any hurry," said Andrew graciously, being quite elated at the prospect of having revenge.

" Come along, then, you little moppet "—Elea nor was fully as tall as himself—" you will have your own way."

" Not always, father," said Eleanor, looking back with a bright smile as she led the way to her own *boudoir*.

" Now, father," said she, when he had looked over the drawings and given his due meed of praise to their execution—" now, father, you pro-mised to grant me any favor I might ask if I would only stay at home."

" Yes, but I'll never forgive Dorothy for asking you," interrupted Ousely, in a gruff tone; " she had no business to do it. How the d—l does she think I could live a whole month without my little Eleanor, eh ?"

" Well, that is not the point in question, father

you must forgive my aunt Ormsby, for you know she, too, loves your Eleanor dearly, almost as dearly as her father does," and she put her arms coaxingly around his neck. "I have given up the pleasure of paying my aunt a visit, and you promised to grant me a favor. Now I am going to ask one."

"And what may it be ?"

"Only to quash these proceedings against O'Daly; nothing can be more absurd than bringing such an affair into court, and it will be sure to do more harm than good to *the cause;* and then the O'Dalys are so much respected, and they are in such distress, that the sympathy of the people will be strongly aroused in their favor."

"And who the d—l cares whether it is or not ?" cried Ousely, his ire beginning to rise.

"I know, my dear father, I know," said Eleanor in her sweetest accents, "but then I have set my heart on this matter, and you will not refuse my request, more especially as your promise is. at stake. If you do, I shall think you don't love me."

"Then you'd think what isn't true, Nell. There, d—n it, I'll not refuse you, only tell McGilligan yourself, for he'll be d—d disappointed."

"Leave all that to me, my dear father," said Eleanor, still preserving her tranquillity of tone and manner. "I'll take 'ः upon myself to dismiss the plaintiff in this case," and she smiled.

"Thanks—a thousand thanks, my dear father;" she stooped, for he was sitting, and kissed his forehead. "So now you can go wherever you like, and I will return to the breakfast parlor. I suppose my mother is pretty well tired of Andrew by this time," she said to herself as she tripped down the grand stairs and along the hall. The squire decamped through a side door, whistling "The Protestant Boys."

"Mr. McGilligan," said Eleanor as she entered the parlor, "my father bids me say that he has come to the conclusion that you had better drop this affair; he is sensible now that no good could come of it."

"How, Miss Ousely!" said the Bible reader, slowly, fixing his leaden eye upon the young lady's face. "That was not his opinion when he left here a few minutes ago."

"I grant you it was not," replied Miss Ousely, "and I will further admit that it was I who reasoned him into this conviction; but I intend to indemnify you for my share in your disappointment. You know you have frequently asked me to visit Mr. Jenkinson's and Miss Gregory's schools, and I have never yet done so. I will go to-morrow and examine the children with one or two of my friends—will that suffice?"

"Oh, surely miss, surely." This was a favorite

word with Andrew. "It's hard, though, that couldn't get those rascals punished."

"Punished for what, Mr. McGilligan?" asked Eleanor with an arch smile. "But, at all events, a bargain is a bargain; you give up your suit and I give up my aversion to visiting schools—all fair, you know. Mother, of course you'll go with me, as you often go alone."

"With great pleasure, my dear, and I am truly rejoiced to hear that you propose going; it is what you should have done long since. I am quite sure, Mr. McGilligan, that my daughter's appearance, and her beginning to take an active part in our affairs, will do a great deal of good."

"More than the prosecution would, at all events," added Eleanor.

"It may be so," muttered Andrew, who was still far from being satisfied, but be dared not persist any more, fearful of losing even the ground he had gained.

"And now you'll be kind enough to leave us to ourselves," said Eleanor, seeing that the Bible reader manifested no intention of moving. "My mother and I have something particular to do; you must therefore excuse us."

"Oh, surely, miss, surely; I hope you'll not forget your promise, though, of visiting the schools."

"I seldom *do* forget a promise," said Eleanor, with a quiet dignity that well became her. "Good morning, Mr. McGilligan!" then ringing the bell, she ordered the servant who appeared to show the gentleman to the door.

" Well, Eleanor," exclaimed her mother, as the door closed on McGilligan, " you have a strange way of your own. How in the world can you treat people so cavalierly ?"

" Why, mother, that is the only way in which you *can* get rid of such people. With all due def. erence to you, your Scripture-reader, or tract-vender, is about the greatest bore in creation. De-fend me from giving encouragement to such gentry. But it is time I was making my toilet. Do you go 'to Clareview this forenoon, mother ?"

" No, my dear, I think not," said Mrs. Ousely, as she drew her chair still nearer the fire, and placed her feet on the fender ; " the weather begins to be chilly, and my blood is not as warm as it used to be. You must ride over alone, except you can get your father to go."

" Oh, I can easily manage that; I have gained a greater victory than that this morning."

" Ah, that's true," said the mother ; " I was for-getting to ask how you managed to carry your point."

Eleanor told her mother in a very few words how she had overcome her father's obduracy, and they were still talking the subject over when the servant partially opened the door—"Is the mistress in there, miss ?"

"Yes, Anne ; what's the matter ?"

"Here's Tom Malone, ma'am, wanting to see you."

"Let him come in, then."

Tom was ushered in accordingly. He was a thin-faced, under-sized man, with a shrewd, knowing look, but his habiliments were in a sad state of dilapidation, and he was otherwise the very picture of a man by whom dame Fortune had dealt unkindly. He carried in one hand an old battered *caubeen*, and in the other a stick, which supported his tottering limbs, for, though scarcely arrived at middle age, poor Tom Malone was infirm and well nigh helpless.

"Your sarvint, ladies," said Tom, as the servant closed the door. "I'm sorry for troublin' you so early, but I was afeerd you'd be out if I'd wait any longer. I wanted to spake to the misthress regardin' a little business of my own."

"And what may that be, Tom ?" said Mrs. Ousely, in a kind tone, while Eleanor prepared to leave the room, seeing that her presence was not required.

"I hope you'll not be offended at me, Mrs. Ousely," said Tom, "for only I couldn't help it, I'd never make free to throuble you."

Why, Tom, if it be any help you want, there

ts no need for your making an apology, it is not the first time you have asked charity of me."

"True for you, ma'am," said Tom, "an' it's myself that always found you an' Miss Eleanor here the kind, good friends, may the Lord give you the worth of your goodness to me an' mine. But it's not *that* that brought me now, misthress dear, only to talk to you about the little girl, ma'am."

" Who, Nancy ?"

" Yis, ma'am, it's about Nancy I came this time." Eleanor turned back from the door, and sat down to hear what would follow. "I'm tould, ma'am, that you're wantin' Nancy to go to church, an' I jist made it my business to come up an' see if it's thrue, for my mind is greatly throubled ever since I heard it, which was only last night."

Eleanor looked at her mother and smiled maliciously. Mrs. Ousely blushed slightly, but she answered quickly : "I have never attempted to force any of my servants in that respect, but I *do* occasionally advise them, for their own good. *I have* spoken to Nancy sometimes on the subject of religion, but as yet I have not succeeded in convincing her. Poor Nancy is very ignorant, I must say."

" An' so is her father, too, ma'am. We're both poor ignorant creatures God help us ! but then our ignorance won't hindher us from gettin' to heaven, if we only do what the Church and the clargy tells us to do."

"But how do you know that, Tom ?" demanded Eleanor earnestly. "You confess yourself ignorant, how then can you be sure that you are in the right road to heaven ?"

"Why, bless my soul, Miss Eleanor! there's no need of larnin' to know that. I know my catechism well, thanks be to God, an' I'm as sure that I'm in the right way as that I'm sittin' here this minnit. I'd rayther than a good deal that you could say the same, Miss Eleanor, an' the misthress, God's blessin' be about you both." Eleanor sighed, and smiled faintly, but said nothing.

"Well! but about your daughter, Tom !" said Mrs. Ousely. "I hope you do not think of taking her away."

"Indeed an' I do, ma'am, beggin' your ladyship's pardon, that's jist what I came for, if it's plasin' to you."

"But it is not pleasing to me, for I feel a real interest in the girl, because of her simplicity and goodness of heart. If you will allow her to remain, I promise you that I will do all I can for her."

"Yes, ma'am, but you could only do for her body, an' sure that's not the main thing, at all. Now, Mrs. Ousely, ma'am, the short an' the long of it is this. If I could let her go to hell with any

one, it 'ud be with you, but you know I can't do
that, ma'am, at all, at all, for you see God gave
her to me to bring her safe to heaven, an' if I
didn't do that, but let her go headlong down into
the bottomless pit, how could I face Him, or what
could I say when He'd ask me, 'Tom Malone!
where's that little girl I gave you?' Oh, bedad,
ma'am, that would never do at all, so with your
lave I'll take Nancy home with me, an' we had
best be off before the masther comes, or he'll be
ragin' mad, an' there's no use, ma'am, in puttin
him in a passion. Maybe you'd be good enough,
Miss Eleanor dear, to ordher the sarvints to send
Nancy home with me, an' to bring her duds along
with her."

Eleanor could scarcely speak for laughing, and
nodding assent to Tom, she asked her mother how
she liked Tom's logic. "For my part," said she,
half jestingly, but really in sober earnest, "for my
part, I consider it unanswerable. Shall I go, my
dear mother, and order Nancy up?"

"Just wait a moment, Eleanor! Now, Tom, I
am really sorry to part with Nancy, will you not
let her stay, if I pledge you my honor that I will
never again say a word to her about religion?"

"I'm heart sorry, ma'am, to have to refuse you,
but I can't do it, at all. It's an ould sayin' that
*there's many ways to kill a dog besides chokin' him
with butter,* an' so it is with the religion, ma'am

Even if you wouldn't say anything to her about it,
there's the masther, that couldn't keep from it if he
was paid for it, an' then there's always a pack of them
Bible-readers an' Jumpers, an' the devil knows what,
back and for'ards to the house, so it isn't safe
quarthers for an innocent little girl like my Nancy,
that's not able to deal with the schamin villains—
I ax your pardon, ma'am, an' yours, Miss Eleanor!"

Eleanor turned to the window to conceal the
smile which she could not repress, but her mother
frowned, and began to look very coldly on poor
Tom. "You can go, then, and take your daughter,"
she said stiffly, "and you need not trouble yourself
coming up to the Hall again. I cannot encourage
a person who speaks so uncharitably of God's
faithful servants."

"Well! ma'am, it can't be helped," said Tom,
grasping his stick and his caubeen, "I'm thankful to
you, your ladyship, for what's past, except wantin'
to turn Nancy, an' all I can do for you I will, that is,
pray to God to bring you into the right way.
God be with you, ma'am, an' you too, Miss
Eleanor, may His blessin' be about you every day
you rise." So away marched Tom, muttering in
an audible voice, "God's faithful sarvints, *iraugh!*
- -faix, it's not God they're sarvin' any how—I'm
thinkin' it's the ould gintleman below! but what's
that to me, so long as Nancy's kept clear o' them!"

In a few minutes Nancy, a neat, tidy girl of

sixteen or seventeen, came into the room with her little bundle in her hand, to bid the ladies "good bye," and to thank them for all their kindness to her.

"Well, Nancy," said her mistress, "this is very sudden. Are you not sorry to leave your place?"

"I am indeed, ma'am, for God knows both you an' Miss Eleanor have been as good to me as heart could wish, an' I'll never get sich a place again— I know that well, but then my father says I must go, an', of coorse, I must when he says so."

Both the ladies expressed their regret to lose Nancy, and although her wages were all paid up before-hand, yet Eleanor placed a crown piece in her hand as she closed the door after her.

"Now, mother," said the young lady, looking in for a moment before she went up to dress, "what do you think of all this?"

"Think of it?" replied her mother; "why what can I think of it, only that these poor benighted Papists are exceedingly obstinate!"

"Ah mother! mother!" and Eleanor held up her finger in playful admonition, "ah! mother! is it obstinacy, or constancy—which? I much fear that they are more to be respected for resisting than we for attacking. But, mercy on me! mother, what an hour it is. I must be off at once!"

"Stay, Eleanor, my dear! I'll be up stairs with you— on second thoughts, I will go with you to

Clareview. You can tell Ben to have the phaeton
at the door in half an hour."

The phaeton was brought round at the appointed
time, and the ladies were just stepping in, when
Mr. Ousely issued from the covert, his fowling
piece on his arm, and Prince at his heels. "Hillo!
Hetty! Nell! where are you bound for now?"

"For Clareview, my dear," replied Mrs. Ousely in her quiet tone, "will you join us?"

"No, by ——, Hetty! you'll not catch me in haste
there again! The priest and I shall never meet
again at Dixon's table. What the mischief brings
you there, either of you, when you don't know but
it's some Popish people you'll meet."

"And if we do, my dear father," said Eleanor
gaily, "I hope you are not afraid of us. None of
them will try to convert any of *your* family."

"Right again, Nell, by Jove! they know a trick
worth two of it. Drive on, Ben;—mind how you
handle the reins, my lad."

"Oh! never fear, your honor, never fear. I'm
jist the boy to take good care o' the ladies. Is it
to Clareview, ma'am?"

"Yes, Ben, and Miss Ousely wants you to stop
at Hampton House as we pass."

"I say, Hetty," called out Mr. Ousely; "you'll
be back to dinner, won't you?"

"Certainly, my dear, we shall not stay very long.
Good-bye."

The ladies had gone but a little way, when whom should they meet but Jenkinson, the school-master, going up, full speed, to the Hall. Taking off his hat very politely to the ladies, as they passed, he asked Ben if Mr. Ousely was at home. Whether Ben heard him or not, he drove on without making any answer; but as soon as Jenkinson was left a little behind, he said to himself, loud enough to be heard, "The devil send you knowledge, you sour faced hypocrite."

Eleanor looked at her mother, but the good lady was, or appeared to be, wrapped in her own reflections, so her daughter contented herself with saying internally, "Worse and worse; even Ben, Protestant as he is, has no love for the Jumpers."

Having made a short visit at Hampton House, and a longer one at Clareview, the ladies set out on their return, and had nearly reached the gate house, when they perceived our old acquaintance, Phil Maguire, marching up the avenue before them, in company with one of their own servants.

"How do you do, Mr. Maguire?" said Eleanor, in her calm, sweet voice, as the carriage passed him.

"Why, blood alive, Miss Eleanor!" said Phil, putting his hand respectfully to his hat, and evidently well pleased at the meeting. "Is it here have you—and the mistress too—your sarvin. ma'am!"

10

" Wh.it business have you on hands now, Mr Magui e, that you're going up to the Hall to-day?"

" Bad 'cess to the bit, Miss El anor, but the masther sent for me, so he did, an' myself doesn't know the raison, except maybe a guess or so. This dacent boy tells me that Jenkinson the school-master's up at the house, an' maybe it's him that wants to see me, for I know he has a mighty great regard for me, an' especially since this mornin'—heth, he didn't lose any time," he added, in an under tone.

" Why, what took place this morning, Mr. Maguire ?" demanded the young lady, with a smile.

" Oh, nothing, Miss Eleanor—nothing at all, only a couple of children that I took from the school below. But I'm right glad to see that you're goin' home, Miss, for I want you to spake a soft word to the masther for me ; not that I care much, one way or the other ; but then a body doesn't like to be abused for nothing at all, an' besides, I'm afeard o' my life that I might lose my temper, an' say something that the masther wouldn't like to hear, if he'd be comin' too hard on myself or my religion."

" Never mind, Mr. Maguire," said Eleanor, laughingly, " I'll be on the spot, and you may depend upon my best exertions being used to keep the peace."

"Long life to you, Miss Eleanor, and many thanks!—I'll get *somebody* to thank you, besides myself!" added Phil, with a knowing look.

Eleanor blushed and smiled, and told Ben to drive on, pretending not to have heard Phil's last words. She did, however, and so did her mother, who raised her veil, and looked sharply at her daughter. "What on earth does he mean, Eleanor?" she said, in a low voice; "for I see by your face that you understand the allusion." Eleanor put it off with a laugh, observing that every one knew Phil Maguire and his droll way of talking. Mrs. Ousely shook her head, but said nothing.

When they reached the house, they found a consultation going on between Jenkinson and Mr. Ousely, the latter from time to time raising his voice to give vent to a thundering oath, or a violent imprecation against Popery. Eleanor looked in as she passed, but her mother went straight up stairs, having no wish to interfere in the impending quarrel.

"Hard at work, father!" said Eleanor, laughing,—"hammering away at poor Popery. I wouldn't be Father O'Driscoll now, for a good deal."

"Are you coming to help us, Nell?"

"Not just yet, father. I'll be back in a few minutes." And away she tripped, to join her mother in her dressing-room.

CHAPTER VI.

Oh! woman, lovely woman! nature made thee
To temper man; we had been brutes without thee:—OTWAY

"ELEANOR, my dear!" said Mrs. Ousely, as she threw herself on a lounge in the dressing-room, "I would not mind, if I were you, going down to the parlor just now. I do not see what interest you can have in listening to those tiresome disputes!"

"My dearest mother!" cried Eleanor, who was already at the door, on her way down stairs, "I wouldn't miss that scene for anything. There's a volume of *Madame de Sevigne's Letters* just by you, that will amuse you till I return." And without waiting for an answer, she descended to the back parlor, where she found Phil Maguire just establishing his burly person on a chair near the door, and listening with imperturbable gravity to a most abusive harangue from Mr. Ousely. He merely looked round as Eleanor entered, without even turning his head. Ever and anon he glanced at Jenkinson, who sat staring at Ousely with mouth and eyes open, greedily drinking in his words.

"And now what have you to say for yourself Maguire?" said Ousely, by way of winding up.

" My answer's very short, your honor," replied
Phil, without moving a muscle, " that I'd do the same
over again the night before the morrow, if there
was any necessity for it. That's jist what I have
to say, Misther Ousely! an' if it's not plasin' to
you, I can't help it. I'm a man that never goes
round the bush to tell what I think."

"What!" cried Ousely in a raised voice, "do
you mean to insult me in my own house?"

" I'd rayther cut my tongue out, your honor,
than insult any gentleman in his own house, an'
you least of all. I only answered the question you
put to me, an' I meant no offence to any one."

" Blood and furies, man! what business had you
to concern yourself about the brats? what was it
to you where they went to school?"

" Not much, to be sure," said Phil coolly, " only
that I knew their poor father, God rest his soul!
an' I knew him for the heart an' soul of a good
Catholic—so is their mother, too. She's workin'
at home with my woman—so you see, your honor,
I couldn't see the children goin' to the devil, knowin'
what they ought to be!"

" Going to the devil!" repeated Ousely, his face
rimson with rage, "how dare you speak so to me?"

" Faith, an' I don't know where else they'd be
goin', if they'd be left in the hands of sich lads as
him," pointing to Jenkinson. " Didn't the woman
kidnap them off the road-side?"

"Why do you speak so disrespectfully of the lady, you wretched man?" said Jenkinson, in his deep, solemn tones.

"God forbid I was as wretched as you are!" retorted Phil. "As for 'the lady,' as you call her, I'll respect her as a lady when she acts like one. It isn't very seemly conduct for a lady to be tellin' lies, an' hoodwinkin' poor simple children, an' inveiglin' them into the den where you an' the likes of you's doin' the devil's work!"

"By the Lord Harry! Phil Maguire," shouted Ousely, jumping to his feet, "I'll make you sorry for this—I'll—I'll turn you out on the road, by ——."

"You forget the bit of a lease, Mr. Ousely!" returned Phil very composedly. "Thank God! I had it secured before you lost your senses—I beg your honor's pardon—I mane before the black gentry got about you—if I hadn't I might whistle for it now, I'm thinkin'."

"Mr. Ousely!" said Jenkinson, his thin lips trembling with anger, "Mr. Ousely! is there no law to punish such a villain?"

"I tell you again," said Phil, taking the word out of Ousely's mouth, "I tell you again, my good Bible-reader! not to be callin' sich hard names!—except yourselves, an' no one expects the thruth from you, there isn't man, woman, or child for miles around Lough Corrib that would call Phil Maguire a

villain !   As for the law, you may do your best.
I havn't done anything to make me afeard of it.
You gave me the throuble of comin' up here, an' I
tould you before you done it, that I didn't regard
any man in fair play.   You depend on havin' his
honor here to back you up in all your schamin'
villany—more shame for his father's son to have
anything to do with you—but I tell you again, here
before his face, that while breath's in *my* body, I'll
never see a fatherless or motherless child that I
know ought to be a Catholic, inveigled into your
school, but I'll have it out, or I'll know for what.
D'ye hear that now, Masther Jenkinson?"

"Scoundrel!" cried Ousely, laying hold of Jenkinson's walking-stick, and brandishing it furiously,
"Scoundrel ! I'll teach you to respect your betters
—by —— I'll break your head, thick as it is!"

"Do, your honor," said the imperturbable Phil,
standing up, however, and placing himself on the
defensive. "Do, strike a man in your own house,
an' you sent for him on business. That will jist
crown your charackter."

But all Phil's rhetoric would not have prevented
Ousely from striking him, had not Eleanor laid
hold of the stick.   Her father turned quickly to
see who dared take such a liberty, for he had
forgotten Eleanor's presence, and on seeing his
daughter, exclaimed : "What the d—l do you
mean, girl ?   Let go the stick, I say !"

" No, father, you must excuse me," said Eleanor
quietly, though her cheek was covered with the
burning blush of shame. "I will not let go the
stick for such a purpose. Give it to me, father
you will thank me for this hereafter."

" Get away with you, girl!" said the father, but
he gave up the stick, "why do you interfere? Am
I to suffer myself and my friends to be insulted by
every clown who chooses to forget the respect due
to gentlemen? D——n it, Nell! give me back
the stick!"

" By your leave, father, I will rather send it
beyond your reach and mine!" She approached
the open window, and sent the stick flying far over
into the wood.

" Miss Ousely!" said Jenkinson, "I beg to
remind you that that stick is mine, and I value
it highly."

" In that case, sir," returned the young lady with
a winning smile, " you can have it by walking out
into the wood—a little exercise is good for the
health, you know." Jenkinson looked sullen, but
Ousely could not refrain from echoing Phil Maguire's hearty laugh.

" Well done, Miss Eleanor," said the worthy
farmer, his honest face glowing with satisfaction,
" By the laws, that was well handled. I'm sure it
would go hard with me when I'd raise hand or foot
against the masther, afther all, but *you* saved me

the throuble this time, long life to you, miss, an
that you may never get the foolish notion into your
head of makin' Protestants out of l'apists. That's
as good a wish as I can make for you now, when
most o' the quality round here are goin' mad about
it." Eleanor thanked him for his good wish.

"Keep your prate to yourself," said Ousely;
"we've had too much of it already."

"May I go, then, Mr. Ousely?" demanded Phil
with comical gravity.

"You may, an' be d——d to you, but mind, if
I ever have it in my power, I'll pay you for this.
We'll see when rent-day comes, whether you'll be
on the same tune. It will soon come now!"

"Let it come when it may, your honor, I'm
ready for it, thanks be to God! I've your rent
ready for you in Bank of Ireland notes. It isn't
poor Bernard O'Daly that's in it, Mr. Ousely,
awow!"

"Be off out of my sight, then," cried Ousely,
"or I may still be tempted to do what I wouldn't
wish to do, in this house, at least!"

"Thrue for your honor—good bye, Miss Eleanor!" and he bowed respectfully—"may the Lord
protect you this day and forever more, amen!
The back o' my hand to you, Misther Jenkinson!
you made a poor fight of it, afther all—my soul
to glory, but you did!" Then making another
low bow to Mr. Ousely, he opened the door and

walked out with as independent an air as if he were the master of the house.

"What a d——d sturdy old fellow he is!" said Ousely, looking after him; "now there's a man you can make nothing of, he's just as unbending as an old oak."

"A true emblem of his religion, father," said Eleanor. "I much fear that after all the expenditure and all the trouble, the people of Ireland, take them as a body, will be just as Catholic—nay, do not frown so, dear father, I mean to say, as Popish as they now are, though you keep at them these fifty years to come!"

"You take good care, Eleanor!" said her father testily, "that *you* don't fail in your efforts, for you make none; by the Lord Harry! you're a d——d queer girl, and I scarce know what to make of you!"

"My dear Miss Ousely!" said Jenkinson, in his smooth oily tones, pulling up his shirt collar at the same time, "My dear Miss Ousely! it is dangerous, exceedingly dangerous to entertain, or manifest any sort of sympathy for these Romanists—they deserve none, my dear young lady!"

"Indeed, Mr. Jenkinson?" said Eleanor with ironical emphasis, "why really now I *did* believe that there were *some* good people amongst the Catholics, and that they *had* a sort of claim on our compassion and sympathy, inasmuch as that the hand of

the Lord weighs heavily upon them just now. But
possibly I might have been mistaken, and I feel
grateful to you, my good sir, for enlightening me!"

Ousely looked from one to the other, scarcely
knowing whether Eleanor was serious or not, but
Jenkinson himself was keenly alive to the piercing
irony of the young lady's tone, still more than
that of her words, and he bit his thin pale lip until
it actually assumed a roseate hue, and, not daring
to make any show of resentment, he fidgeted on
his seat, and muttered something about the evening drawing on.

"Why what the deuce hurry are you in?" cried
Ousely, who was already busily engaged examining
some new fishing-tackle. "What did Nell say to you
that you look so blue on it—eh, Jenky—d— it, look
up, man! there's money bid for you!"

"You are very good, sir, but—I know not how
it is—Miss Eleanor! your words are not what I
would expect"—he shook his head. "Ah, my dear
young lady! *out of the fulness of the heart the
mouth speaketh*—so it is written!"

"What the d—!, Nell—" began her father.

"Never mind, my dear father!—worthy Mr.
Jenkinson, have no fears for me—I am sound to the
core," and she laughed merrily. "In proof where
of—as some of yourselves would say—I am going to
pay you a visit to-morrow in your *sanctorum*—do
you understand?"

"Oh, of course I do," said the schoolmaster though he looked somewhat puzzled. "Do you indeed propose to honor my poor place with a visit, Miss Ousely?"

"If my dear father has no objection," said Eleanor, turning to him.

"Objection!" cried Ousely, "why, upon my honor and soul, Nell, I'm well pleased to hear you say so—I'll go with you myself, by h——, just to see how the young Papist brood can act and talk like good Protestants."

"You are very kind, sir," said Eleanor, though in her own mind she determined that her father should not be of the party; then turning to Jenkinson, who stood with his hat in his hand, and a cold smile on his sallow features, "You have not seen your excellent friend, McGilligan, this afternoon?"

"No, Miss Ousely," said Jenkinson, with a piercing look, "that devoted Christian has not much time to spare for making visits."

"Of course he has not, Mr. Jenkinson! but I signified to him my intention of visiting your school and that of Miss Gregory to-morrow—you will have the school in readiness, as there are some of my friends going with me, and if you have no objection we shall examine the children. Good morning, Mr. Jenkinson!"

"Good morning, Miss Ousely, I am elevated

and highly honored—yea, far beyond my poor
deserts."

" Oh. you are far too humble, my good sir !"
said Eleanor with her own peculiar smile. " Humility, you know, is a Papist virtue, and would
never do for a Bible Christian, especially a teacher
—always keep up your own dignity, my dear sir,
and have what the pious Scotchman prayed for—
a good opinion o' yoursel'—take my word for it, it
is the best way to ensure success—good morning
once more—father, it is drawing near five o'clock,
dinner will soon be on the table !"

When the door had closed on Eleanor's graceful
form, the two men stood looking at each other in
silence, and it was not for some minutes that
either spoke.

" What a comely young lady your daughter is,
Mister Ousely !" said Jenkinson slowly, " and a
clear-sighted, quick-witted young lady, but"——

" But what, sir ?" demanded Ousely, almost
fiercely.

" Why, dear me ! Mister Ousely, don't be angry.
I was only going to say that I fear she does not
inherit the fervent zeal which moves her father to
do and attempt great things for the good cause—
the Spirit tells me that she has not sought or found
the Lord !"

" Then the Spirit tells you a confounded lie !"
returned Ousely, waxing wroth, " for if Nell

Ousely hasn't found the Lord, as you say, then
don't know who has, by jingc ! my dear fellow,
that girl's worth half a score of your Bible-readers,
and whatever she may choose to say now and then
when she's in the humor for quizzing, I don't believe there's a better Christian or a sounder
Protestant in the country !"

" I'm glad you think so, Mr. Ousely."

" Think so I I tell you, sir, I'm sure of it—that's
your way out—your Spirit tells you, in leed !"

" But, Mr. Ousely, my very good sir !" said
Jenkinson, with unfeigned sorrow, " I can assure
you that I meant nothing, nothing hurtful to Miss
Ousely; far be it from me, sir, to say, or insinuate,
or even think anything but what is good of the
young lady, whose talents and virtues, not to speak
of her exceeding comeliness, are known and
published, and commended all the country over.
I only meant, my honored patron, that it might
be well to advise the young lady, just between
yourselves."

" The d—l you did ! I tell you what it is now,
Jenkinson ! I wouldn't speak to Eleanor on such
a subject, no, not if the success of your missions
depended on that one word. Why, man, my Nell
could give chapter and verse for everything she
does and says, and neither you nor myself could
hold the candle to her—she knows and understands
everything—aye ! everything ! a d——d deal better

than I do, so no more about that, if you want to
keep on good terms with me. Just let Eleanor
alone, and be thankful when she shows any disposition to interest herself in the missions! Here,
John! John!" The servant put in his head.
"Show Mr. Jenkinson out!" Not a word more
would he listen to, and the discomfited school-master could only give vent to his pent-up feelings by
holding up his hands, and shaking his head, as he
passed through the hall with the servant, saying,
as the latter opened the hall-door to let him out,
"They are all touched here, of a surety they are!"
touching his own forehead as he spoke.

"Some of them are, at any rate," replied John,
as he closed the door after him, "or they wouldn't
encourage the likes of you."

When Eleanor joined her mother in the dressing-
room, she proceeded to give her an account of the
scene she had just witnessed. "Now what do you
think, my dear mother," she concluded, "of this
system of kidnapping children, for view it as we
may, it amounts to that?"

"My child," said Mrs. Ousely, "we are not to
view these things with a merely human eye, let us
view them as God views them."

"Well! and what then?" demanded Eleanor
with an arch smile; "are we to suppose that God
can ever sanction fraud in any shape or form!"

"Oh! of course not, my dear, but then, you

know—in short, Eleanor! I can scarcely justify
this particular instance, but I am confident that
Mrs. Perkins meant well, and if ever deceit can
be harmless, it is when practised in the cause of
religion. The whole success of the mission depends on having the training of the rising generation,
and any means are justifiable that tend to secure
that most important end."

"Well, mother! there is little use in arguing
the question now—for my part I have doubts, and
very serious doubts, as to the usefulness of these
schools, and I am not at all prepared to justify
proceedings which I believe, on the contrary, to be
wholly unwarrantable. I have no idea of such
India-rubber consciences, that can be made to suit
any emergency, or rather expediency. However,
here is John—coming to announce dinner, I dare
say!"

"Yes, miss, dinner's on the table, and the master is waiting."

"Come then, mother!" said Eleanor, "take my
arm, and let us go down."

"Mind you don't say anything to your father
about the schools, Eleanor!"

"Oh, never fear, mother! you know I very
seldom speak to my father on such subjects—and
when I come to think of it—what a day we have had
of it with these humbugs—I beg pardon, mother!"

she quickly added, " I mean these worthy people
who follow the godly pursuit of proselytizing—"

" Proselytizing, Eleanor ?"

" Oh ! another slip of the tongue. I mean,
waging war on Popery, naughty giant that he is !
pulling him by the legs and tugging away at the
skirts of his huge coat, and then crying out, might
and main, that they are killing him !" and Eleanor
laughed at the ludicrous image presented to her
mind.

" Dear Eleanor !" said her mother, "how strange-
ly you do talk at times !"

"Why, surely, ma'am, surely, as worthy Master
McGilligan would say—truth is always strange
amongst evangelical people ! Did you never know
that before ? But here we are—so no more at
present," and she opened the dining-room door,
where her father was already growing impatient,
" for those confounded fellows," said he, " have kept
me so busy all day that I have not had time to
take my usual luncheon ! Upon my honor and
word, they're enough to drive a man mad with their
rascally squabbles !"

The ladies smiled assent, and the trio took their
places at the table, where we shall leave them, and
take a glance at Larry Colgan's little domicile, to
see what is going on there. The tall gate-keeper
had just put his spade in its usual place behind the
door, and taking a pipe from his waistcoat-pocket

drew a stool to the fire and sat down. His wife,
a rosy-faced, happy-looking personage, precisely
what Lord Byron would call " a dumpy woman,"
was sitting spinning flax at a little distance, beguil-
ing the time with an old melancholy ditty which
he sang in a low monotonous voice, when the door
opened and in walked our old acquaintance Granny
Mulligan, her bag on her back and her staff in her
hand.

"God save all here !" said the old woman.

"God save you kindly, hònest woman !" returned Peggy Colgan, suspending her employment ;
"won't you come by an' sit down ?   Have you travelled far the day ?"

"Not very far—only over from Bernard
O' Daly's ! But don't you know me ?"   Here
both husband and wife looked more closely, for the
light was already wearing dim, and both started to
their feet ! " Why, then, tare-an-ages ! Granny
Mulligan, is it yourself that's in it—why, woman
dear ! how's every teather's length of you ?"

" Well, I can't complain, Larry ! thanks be to
God for it ! You needn't trouble yourself, Peggy
ahagur, to be puttin' away my bag !"

" No? why then, to be sure you'll stay all night,
granny—of course you must !"

" I wo.ld and welcome, Peggy dear, only I promised, God willin', to be back at Bernard O'Daly's
'fore bed-time—God help them, they're in the

height o' trouble, the craturs! for poor Honora's taken very bad. Still they wouldn't hear o' me goin' anywhere else to lodge!"

" Why, Lord bless me!" said Larry, " it must be very suddenly that Honora was taken sick, for she was up at the Hall this very mornin'."

" I know she was, Larry, an' it was the unlucky journey for her—you know she was in the worst of health this many's the day, but walkin' so far this mornin', and the way that Misther Ousely spoke to her—just as if she was a dog—an' worst of all his offerin' to give a clear quittance if the family would turn Prodestan'."

" An' did the masther say that to her!" demanded Larry, taking the pipe from his mouth, and holding it suspended between his finger and thumb. " Did he mention the likes o' that to Mrs. O'Daly ?"

" Indeed then he did, Larry *acushla !*"

" Well, by my sowkins!" cried the gate-keeper with honest indignation, " that's the hardest thing he done yet. Oh, now that bates Banagher, and the devil to boot. I knew very well that he'd do nothing for her, an' that may be he'd tell her the raison plainly—bekase of the family bein' sich good warm Catholics—but I hadn't the laste notice that he'd hint sich a thing as turnin' to her or hers. Oh! then, Peggy dear! was'nt that a hard turn to do ?"

"Hard!" said Peggy, "why I tell you there's nothing too hard or too hot or too heavy for them Bible-readers, for it's them that's puttin' all this in the masther's head. Sure don't we all remember when he wasn't half so bad as he is now—he'd give you a rally, to be sure, if you hadn't the rent with you, but then it was soon over, an' he'd give you time, an' he wouldn't make a bit difference atween' Prodestan' an' Catholic, but from the day he got in with that thievin' crew he wasn't the same man. Och, indeed, but it's myself that's heart sorry for poor Honora O'Daly, an' good right I have, for, before Larry an' myself struck up together, I lived for two years on her flure, an' a better misthress I never sarved a day to."

"An' then she had sich a pride out of her family," observed granny, "bekase they were so good an' so pious, an' them so comfortable about only a year or two back." Granny stopped to wipe her eyes, an' Peggy's were not dry. "You see," continued the old woman, "she thought it 'ud soften the masther's heart when she crept up to the Hall, an' her so feeble, an' so she stole out afore any o' the men was stirrin' in the mornin'—but when herself an' Bridget came back"—

"I sent them back in Jack Connor's cart, that I got a loan of," interposed Larry, "but I thought it was only tired that the poor woman was!"

"I know.—I know.—*acushla!*—well! as I was

sayin', when they came back, poor Honora had to go to bed, au' the whole was found out, an' there's black trouble in the house, for they all know well that Honora will never stand on green grass."

"God comfort them this night!" ejaculated Peg. gy, as she arose and busied herself about the supper. "At any rate, granny, you'll stay an' have some supper with us."

"Why, then, I will, Peggy *astore!* an' thank you kindly for the offer ; to tell you the truth, I came out on the intention of takin' my supper wherever I'd be first asked, for though I'm as welcome now in Bernard O'Daly's corner, an' to take my share of what's goin' as ever I was, still I can't bring myself to do it, for I know—och! och! I do— that there's far from bein' plenty even for them. selves. But still I'll go back, plase God, for I want to sit up with Honora the night."

"Well! but you'll stop a night or two with us afore you go, won't you, granny ?"

"Oh! indeed then I will, an' glad to be asked, it's not every house that we're invited to lodge in now-a-days."

"That's thrue enough, granny," said Larry. "but never mind, there's enough o' the ould stock left yet to give you a hearty welcome wherever you go. Come now, let us fall to !"

When supper was over, the kind-hearted Peggy put a good dish of oatmeal into granny's bag

" Now what are you doin' that for, woman dear !"
said the beggarwoman, as she snatched at the bag,
"I didn't want to take anything from you, afther
gettin' my good supper."

"Well, but listen here, *avich !*" and Peggy
whispered her words into granny's ear, for fear
the children should overhear her ; " you say there's
.scarcity, an' I know there is, where there ought to
be full an' plenty ; now, I wouldn't affront them by
sendin' anything to them, but can't you jist watch
your opportunity, when none o' them's lookin', an
put this with their meal, wherever they keep it
Can't you do that now ?"

"Ay ! indeed can I, Peggy ! God bless you,
zcushla machree ! for the kind, good thought, an'
may He increase your store !"

Larry pretended to be wholly engrossed with
his pipe, and never turned his head till the old
woman bid him ' good night.'

" Good night, an' God be with you, granny ;
you'll be over the morrow, won't you ? but at any
rate, Peggy or myself will take a race down to
see poor Mrs. O'Daly !"

" Declining Larry's offer of going a piece with
her, granny Mulligan grasped her oaken cudgel
and stept out into the darkness, for the night had
closed in, dark and moonless. When Larry had
closed the gate behind her, and returned into his
lodge, an uncomfortable sense of utter loneliness

began to steal over the sturdy old woman, and, for the first time in her life, she felt something like fear. Not that granny Mulligan was afraid of the surrounding darkness, or of any bodily evil befalling her, but still there was a chill creeping over her, and though she battled bravely against it, there was no getting rid of it, do what she would. The whole secret was, that she had to pass by a burying-ground on her way to Bernard O'Daly's, and what made the matter worse, it was the burial ground belonging to the Episcopal Church. "Now if it was our own sort that was in it," said she to herself, "I wouldn't be much afeard, for them that's anointed with the holy oil, an' gets the rites o' the Church at their last hour, won't do any one harm, even if God allowed them to come back again, to get any little matter settled that might be troublin' them, but then it's a different thing to pass by where the Prodestans are, sich a night as this, without a livin' sowl with me. Bedad, I'll go back and get Larry to come with me, afther all."

She was just turning on her heel to go back, when she heard the avenue gate open, and then the pit-pat of more than one pair of small feet, and to her great joy she heard Larry Colgan's eldest boy calling to her.

"Here I am, Thady *astore!* what's wrong with you, *avick?*"

"Nothing at all, granny," said the boy, coming up close to her on one side, while his younger brother caught hold of her cloak on the other "nothing at all, only daddy an' mammy sent Peter an' me afther you, for fear you'd be lonesome, an' bekase you had to pass the Prodestan' grave-yard."

"Hut, tut!" said granny, affecting great bravery, "what harm would them that's in it do me? I never done them any harm."

"Oh! but some o' them might appear to you, you know, an' if you'd see any o' *them*, you'd never get over the fright; dear knows but they'd kill you; them Prodestan' ghosts are evil sperits, mammy says!"

"Well I know that, *avick!* but how are you an' Peter to get back?"

"Oh! daddy's comin' down to see how Mrs. O'Daly is, an' we're to stay there till he comes."

"Well! if that's the case, children, let us go on in the name of God."

For a few minutes they walked on without speaking, but the children were awed by the deep silence and the darksome night, and Thady begged of granny to tell them a story, 'jist to pass the time.'

So granny began the story, nothing loth to hear herself talk, and the tedium of the road was thus beguiled, till the moon began to peep from

behind a gauzy cloud, and her first beam.s glittered
on the spire of the Church, now in sight. '

"God grant us the light of heaven!" said gran
ny; "there's the moon, an' a purty bright moon she
is."

"Ay! but there's the grave-yard, too," said the
elder boy, and both the children clung close and
closer to the old woman. "Look how the white
headstones are standin' up, jist like ghosts."

"Don't be afeard, children, don't be afeard—
bless yourselves now, and then you needn't fear all
the divils in hell."

On they went, and still the spectral-looking
grave-stones grew whiter and whiter in the moon
light, and the shade of the old yew-trees inside the
wall fell deeper and darker across the graves.

"If a body could only pray for them," said
granny with a sigh, "there would be some comfort
in that; but, ochone! isn't it a lonesome thing to
lie there without one to offer up a prayer for
them, an' every body afeard o' their lives to pass
them by? Livin' or dead, children, it's a poor
thing to be a Prodestan'."

"Granny dear!" whispered Peter, "do you see
anything at the gate there?"

"No, I don't," said granny, though her own
voice trembled; "don't you see it's the shadow o
the tree agin the gate pier. What did you think
it was, agrah?"

13

"Why I thought it might be the ghost of Tom Connor, the Jumper. They say he appears in the shape of a big black dog, with fire comin' out of his mouth an' eyes."

"The Lord save us!" said granny, making the sign of the cross devoutly on her forehead. "An' did Tom die a Prodestan'?"

"He did, an' was buried here about two months agone."

"Avoch! avoch! but it's jist what I'd expect from him, he was ever an' always a bad mimber. God knows but the counthry was well rid of him when he went—thanks be to God, it's only him an' the likes of him that dies sich a death as that. But now we're past the grave-yard, children, an' we'll soon be at the house. Step out now, for I'm a long time away, an' maybe I'm a wantin' before this."

CHAPTER VII.

Learning, that cobweb of the brain,
Profane, erroneous, and vain :
A trade of knowledge, as replete
As others are with fraud and cheat ·
An art t' encumber gifts and wit,
And render both for nothing fit.—BUT. SE's Hudibras

MRS. OUSELY and her daughter set out on the following morning to visit the schools, according to promise, and by a little exertion of Eleanor's tact, her father remained at home. When the carriage reached the gate-lodge, Larry Colgan stood ready with his huge key, and when he had thrown the gate wide open, he sidled up to Eleanor.

" Good morning, Larry !" said the young lady, while her mother nodded and smiled ; " I hope you are all in good health here?"

" All well, thank God an' you, Miss—bad manners t) you, Ben ! can't you take it easy—what a hurry you're in this morning ! If you're not in too great a hurry, Miss Eleanor, I'd be makin' free to ask you to stop a minnit—there's a person inside that wants to spake to you."

" To me, Larry ?"

" Yes, Miss, to you, if it's plasin' to you."

" But wha, does the person want with me—why not come out and speak to me here ?"

Larry came up close to the carriage, and said in a low voice : " Sure it's Kathleen O'Daly that wants to see you—she dar'nt go up to the house for fear of meetin' the master."

" Oh! if that be all," said Eleanor, smiling, " there is no need of secrecy, Larry! My mother is just as much interested about the O'Dalys as I am. It is Kathleen O'Daly, mother."

" Go in. then, my dear daughter, and see what she wants. Poor girl! she need not have concealed herself from me. Make haste, Eleanor, I shall wait for you. I am really anxious to hear how Mrs. O'Daly is this morning."

When Eleanor entered the lodge, she found Peggy Colgan doing her best to comfort poor Kathleen, who sat with her eyes fixed on the door in breathless anxiety, apparently too much intent on her own sad thoughts to pay much attention to Peggy's well-meant truisms. The minute she saw Eleanor, her eyes filled with tears, and starting to her feet, she clasped her hands with convulsive energy : " Ah! I knew you'd come in, Miss Eleanor! I knew you would—may the Lord bless you and protect you from all harm ! My mother's dying, Miss Eleanor, dear ; and we have nothing to comfort her poor weak heart. We can't hide it

any longer, Miss Eleanor; and as none of our
neighbors can do anything for us, I thought I'd
come up and make application, where I knew there
was both the *will* and the *way*."

"My dear Kathleen," said Eleanor, taking her
hand kindly, "I trust your mother is not in such
immediate danger as your fears would make her.
Take courage; she may yet recover."

"Never, never, Miss Eleanor!" said Kathleen,
with a fresh burst of grief, "she is dying—dying.
Oh! indeed she is—and it's she that was the good,
kind mother!"

"Kathleen!" said Eleanor, earnestly, and even
solemnly, "I fear you must blame one whom I am
bound to love and honor. for hurrying on this sad
catastrophe. Tell me, Kathleen, is it not so ?"

"My dear Miss Eleanor, don't trouble yourself
about that," exclaimed Kathleen, with sudden
energy; "my poor mother was declining this many
a long day. Indeed she was, Miss ; and if she *did*
get worse since--since *then*—" she stopped a moment, as if to control her feelings—"No, we don't
blame any one, Miss Eleanor ; we take this new
trial from the hands of God, and we bear it for His
sake. Oh! God forbid," she raised her mild
blue eyes to heaven, "God forbid that we'd owe
any one a spite."

"Well, Kathleen," said Eleanor, wiping away
the tears that would rush out, "you may go home
12*

now--don't be afraid that your mother shall want anything. We are going out for an hour or two; but as soon as I get home, I will see that every-thing needful is sent. How is my little favorite, Eveleen?"

"She is very well, Miss, thanks to you for ask-ing, but there's none of us in greater trouble than she is—poor child; well she may be in trouble —she's going to lose the best friend ever she had, or will have."

"Good bye, then, Kathleen, I'll try to see your mother very soon. Good morning, Peggy, how are the children?"

"In good health, Miss, thanks be to God for it."

When Eleanor rejoined her mother, she related what had passed, and Mrs. Ousely was much shocked to hear that Honora was so ill. Her lip trembled with emotion, as she said,

"Poor Honora! I fear her disease is a broken heart."

"It is nothing else, my dear mother. Now, Ben, drive on, time is passing."

"I thought we were to have had Amelia Dixon with us, Eleanor? Did she not say yesterday that she would go?"

"Yes, mother, she was to meet us at the cross-roads at eleven o'clock. We must make haste, Ben, for Miss Dixon may be waiting." Smack went Ben's whip, and off went the horses at a

brisk trot, but they had gone only a little way
down the road when a horseman dashed up at full
speed, and reined in his prancing steed along side
of the carriage.

"Good morning, Miss Ousely!" said he, in a
voice whose modulated tones bespoke the gentleman. "Oh! your mother here too; good morning,
Mrs. Ousely,. I hope I see you in good health
to-day!"

"Tolerably good, I thank you, Sir James!" said
Mrs. Ousely, leaning forward to shake hands with
the stranger. "Were you going up to the hall,
or where?"

"I am in a charitable mood this morning,"
replied the baronet, as he exchanged a meaning
smile with Eleanor, "so I propose to visit the
schools with you, provided you have no objection."

"Oh! certainly not, Sir James! we shall be
happy to have your company. But where is Amelia? she promised to come, did she not?"

"Oh! as to the promise," said Sir James, with an
arch smile that well became his dark, Spanish-looking
features, "my cousin Amelia has changed her
mind, and deputed me to come in her place. She
craves your pardon, and hopes to see you soon.
Why so serious this morning, Miss Ousely? are
you framing your interrogatories?"

"Not so, Sir James!" said Eleanor, looking up
for the first time, "I will trust to the occasion for

suggesting them—there is inspiration n Mr. Jen
kinson's face!" she added with sly humor. "I
was just thinking of Amelia's message, and
wondering why she changed her mind." There
was a meaning in her words that was not lost upon
Sir James.

"I am sorry, on your account, that my cousin
has not kept her promise," said he with some
bitterness; " but even if she had, it is probable
that you should still have had the present incum-
brance, for the temptation was too great to be
resisted. *You* know how desirous I am of gaining
all possible information concerning this great
movement."

Eleanor raised her eyes again to the young
man's face, and though she spoke not a word, yet
he felt satisfied ; that glance said more than words.

"What a changeful sky is this of yours!" said
the baronet, as he gracefully reined in his impatient
charger to keep beside the carriage, " how beautiful are these sudden transitions from cloud to
sunshine, and how many charms do they not bestow
on the features of the country, lovely and varied
as they are of themselves !"

"Yes!" said Eleanor, " our sky is just the one
to overhang a Celtic nation—there is as much variety in the character of our people, when you
come to study them, as there is in our shifting
firmament. Believe me, you will find many

beautiful virtues and many sterling qualities
amongst the unsophisticated peasantry. They are
a people to be loved, ay! and honored, let their
traducers say what they will!"

"My dear Eleanor!" said her mother, "you
speak warmly. Sir James, with his cool English
reason, must think it strange to hear you talk so
of these poor benighted Irish, who are little deserving of respect, not to say honor, in their
present degraded state. If this great work now
in hands can only be made to succeed, then they
may become respectable; were they only disen-
tangled from the meshes of Popery, we might
have hopes of them! You must excuse my
daughter, Sir James! she is young and enthusias-
tic!"

"The apology is scarcely needed, my dear
madam!" said Sir James, whose eyes were fixed
admiringly on Eleanor's blushing face. "My
English reason is not so *cool* but that I, too, can
admire the truly Celtic virtues of the Irish, and
sympathize with their manifold wrongs! It is
precisely because I *can* and *do* that I am here now.
I have heard and read much that is both good and
bad concerning the Irish people, properly so called;
and I have crossed the channel in order to see and
judge for myself."

"Indeed, Sir James?" exclaimed the elder lady
"why, who would have thought it?"

"I, for one, mother," said Eleanor with a smile,
"I partly guessed as much." The young gentleman smiled, too, and his dark eyes sparkled with
pleasure. He was evidently pleased to find that
Eleanor so far understood him, for he had never
before spoken to her of his object in visiting
Ireland. He had not time to make any reply,
when Eleanor exclaimed: "See, yonder is the
school-house, Sir James! the *Alma mater* of the
Jumpers in these parts! How purely white it is,
something like the whitened sepulchres mentioned
in Scripture, I fear!"

"Eleanor, my dear!" said her mother, in a tone
of reproof.

"I beg its pardon, and yours, my dear mother,"
said Eleanor, laughing, while Sir James turned
his head away, lest Mrs. Ousely should see him
smile; "That is, if the comparison be offensive to
you."

"You are an incorrigible girl," said her mother,
with a faint sigh.

"Call me anything you please, my dear mother,
except a hypocrite."

Just at this moment the carriage stopped in
front of the school-house, and out came the long,
thin visage of Jenkinson, at the door, then his
whole gaunt frame sidled out after it, and with
many a bow, and many a grave smile, he welcomed
his distinguished visitors. He was stepping for

ward to offer his hand to Eleanor, but Sir James
sprang lightly from his horse, and saying, "Excuse
me, sir," he gracefully assisted the ladies to alight.
Jenkinson was half inclined to resent the stranger's
interference, but when he cast a cursory glance
over his tall, commanding figure, aud marked the
dignity of his demeanor, he shrank back into himself, muttering, "Second thoughts are best."

"Will you be good enough to lead the way into
your school-room, Mr. Jenkinson?" said Mrs.
Ousely. "Of course you are prepared to admit
us."

"Oh! certainly, ma'am, certainly; will you con
descend to walk this way?"

"So this is the potentate who holds dominion
here?" said the baronet to Eleanor, in a low voice,
as they walked in side by side.

"Yea, verily, this is the righteous, and evangelical, and popery-hating, and Bible-loving instructor
of youth, placed here as a light amid darkness,"
said Eleanor, imitating Jenkinson's own prolix ver-
biage. "You stare," she added, laughingly. "But
you will soon cease to wonder at the superfluity
of words wherewith I do eulogise our excellent pedagogue. Be silent now, good sir, that you may
hear; for, of a surety, Jenkinson is about to hold
forth."

"Mr. Dalton," said he to his usher, a pale, effe

minate-looking young man, " Mr. Dalton, the boys
have not yet recited their scripture lesson."

" No, sir, they are just preparing it."

" Very good, Mr. Dalton, let us have it now.
Ladies, will you condescend to sit down. Sir," to
Sir James, " will you be pleased to take a seat ?"

The visitors being duly settled in their respective places, the master took his station near Mrs.
Ousely, and the pale-faced usher stepped up on a
sort of dais and commanded the boys to close their
books. The order was instantly obeyed, some of
the poor, starved-looking urchins taking a last peep
before they closed their testaments.

"Now commence," said Dalton. " The fourteenth
chapter and first verse, of John. Peter O'Malley.
you say the first verse."

Peter did say his verse, and the others followed
in turn, until the whole of that mysterious chapter was said; some few of the boys making sad
work of it, but in general they said their verses
correctly. When the lesson was ended, Jenkinson
turned to his visitors, with the air of a man who
expected a compliment. Mrs. Ousely was de-
lighted, and told Mr. Jenkinson that he was doing
more to overthrow Popery, than the whole Bible
Society and Tract Society put together.

" You are very good to say so, Mrs. Ousely,"
said Jenkinson putting on a very modest air

" What do you think, sir ?—I am at a loss, ma'am, for this gentleman's name."

" Sir James Trelawney."

Jenkinson bowed very low.

" I hope you are pleased with the boys, Sir James ?"

" They have said their lesson well," replied the baronet, somewhat drily.

" Oh! but you must hear them examined, in order to judge of the progress they have made Lawrence O'Sullivan."

" Well, sir," said a little chubby-faced boy, about eight years old, as he raised himself to a standing posture.

" What is Popery, Larry ?"

" Popery, sir ?" Larry scratched his head, and kept looking at the boy next him, who said something in a low voice.

" Popery's the great delu—" another look at his neighbor—" the great delusion, sir !"

Larry looked much relieved when the last sylla. ble was out.

" Very well answered," said Jenkinson ; " now tell us what *is* the great delusion—you, Terence Landrigan !"

" It's Popery, sir !" Eleanor and the baronet exchanged smiles.

" Very good indeed. Now, Terence, when you've done so well, just tell us who is Antichrist !"

"The Pope, sir!"

"Right again! and can you tell me who was Luther?"

"Luther, sir? Luther was"—Terence's memory was evidently at fault.

"Go on, you blockhead, who was Luther?"

"The—the—the man of sin, sir!"

"Sit down, sir!" cried Jenkinson angrily. "That's the pope you mean." Eleanor pretended to use her handkerchief, and Sir James maliciously said to Mrs. Ousely, "what a smart lad he is' wonderfully wise for his age!"

"Miles O'Callaghan! stand up there!" Miles was a tall thin lad of some ten or twelve years old. "What was the Inquisition, Miles?"

"A place where good men and women were tortured, and put to death for their religion!"

"Very good indeed, Miles! and who were these good people?"

"Protestants, sir!"

"Many of them Jews!" said Eleanor in a low voice to Sir James, who nodded assent.

"Right, Miles, right. And who put them to death, and burned them up?"

"Priests and monks, sir!"

"Right again, Miles. Well! now can you tell me what is confession?'

"Yes, sir! it is an humble accusation of one's self—" began Miles.

"What are you saying, you stupid fellow?"

"Why, that's what's in the catechism, sir!"

"Yes! in the priest's catechism!" said Jenkinson; then raising his voice higher, "can't you tell me what confession is?"

"Why, sir, I *was* tellin' you, an' you wouldn't let me."

"Sit down! John McSweeny!"

"Sir!"

"Who was Queen Elizabeth?"

"Ould Harry's daughter, sir!'

"Henry the Eighth, you mean!" said Jenkinson sternly.

"Yes, sir!"

"What did she do, John?"

"She ripped open the priests, and out the heads off o' them, sir, an' hunted them out o' the country, sir."

"Hush! John!" said Dalton eagerly; "that's not the answer, you're wrong!"

"Why, that's what I heard my father readin out of a book about her!" said John, boldly.

"Put him down to the foot!" cried Jenkinson his face purple with rage. "It is a hard, and v never-ending, and an arduous task," he added turning to the visitors, "to get these Romish children to learn anything!"

"I do not at all doubt it," replied Eleanor, repressing a smile.

"Will you allow me to ask the boys a few questions, Mr. Jenkinson ?" said Sir James.

"Certainly, sir !" returned the schoolmaster, though he and his subordinate exchanged looks that showed their minds ill at ease. "Stand up all of you, children."

The baronet cast a searching glance over the long lines of anxious little faces before he spoke, and then selecting those who seemed most intelligent, he put a few leading questions on the great truths of religion. Alas! he could get no satisfactory answer, except now and then when memory brought back to some of the older boys the almost forgotten teaching of the priest. Thus Sir James had asked several boys the question, "For what end were we created?" and when, at last, the answer came, "To know, love, and serve God, and to be happy with Him forever," the boy concluded with "That's what our own catechism says, sir !"

"And it says right, my boy !" said Sir James, patting him on the head. "That will do, Mr. Jenkinson! we are but trespassing on your time."

"But will you not hear the boys sing a hymn, sir, before you go ? Those questions which you put to them are not those which we generally ask them, so that they were somewhat put about, but you must hear them sing !" Sir James bowed assent, and the ladies resumed their places.

The hymn was one of thanksgiving for the special favor of being "snatched from the burning," and when it was ended and duly praised, the copy-books &c. were exhibited, and then the visitors were ushered into the female school kept by Miss Gregory, where a similar scene was gone through, only that Eleanor, instead of Sir James, examined the girls, and then, having given some money to the respective teachers to be distributed amongst the children, the ladies were shown to their carriage, and the baronet mounted his 'gallant grey,' nothing loth to effect his escape from the schoolhouse. He had not moved a step, however, when Jenkinson laid his hand on the horse's neck, and said, " A word with *you*, sir, before you go !"

" Well, sir, what is it ?"

" You're from England, sir, as I understand."

" Yes—what then ?"

" I would ask you, Sir James Trelawney, to use your influence, when you return home, in behalf of this most glorious and most interesting work—the conversion of the Irish Papists—which is, or ought to be, exceedingly dear to every philanthropic heart. Oh ! sir, if you are a Christian, you will urge your friends and acquaintances to contribute their mite in support of a cause so important in the eyes of God and man !"

" Be assured I shall make honorable mention of your arduous endeavors," replied Sir James,

13*

evasively. "In the meantime suffer me to follow
the ladies, who are leaving me far behind. Good
morning, sir!"

Away rode Sir James Trelawney, and Jenkinson
stood gazing after him for some minutes, then
slowly turning into the house, he said to himself,
with a heavy sigh: "He is no great friend to us—
that I can see with half an eye. I much fear that
he is a Jesuit in disguise. What a pity that he is
such a noble-looking personage—he may be a Romish bishop for all I know—but then he is too
young—some of those English grandees, I suppose,
who have lately gone over to Rome!" Then going
into the school-room, he called for his large ruler,
and began, by way of revenging his disappointment,
to punish some of the boys who had given Popish
answers to the questions put by Sir James.

Meanwhile, the baronet had overtaken the car
riage, and was asked by Mrs. Ousely, what he
thought of the schools. "Is it not truly encouraging,"
she said, "to see so many Romish children of both
sexes conducted into the fold of truth—"

"Pardon me, madam!" said Trelawney, "I am
far from seeing this matter as you do. I much
fear that instead of getting into the fold of truth,
they are getting out of it. I was grievously disappointed this day, for I find, that so far from being
taught anything solid or useful, they are only filling
their minds with trash—the old stale abuse of Po

pery—as they are made to call the religion of their fathers—which may do them no good, but much harm."

"Well! well! Sir James," said Mrs. Ousely, in a somewhat peevish tone, "I cannot see these things as you and Eleanor see them—I, at least, have no leaning towards Popery, that might bias my judgment—I see matters as they really are."

"Yes, but you look through old Protestant spectacles, my dear mother! There, you've a pair of them on at this present moment, which are at least a hundred years old. Those old *ascendancy* glasses, Sir James, are an heir-loom in my mother's family, and came to her from an excellent old uncle and aunt who brought her up."

Trelawney smiled, but said nothing, not knowing how Mrs. Ousely might take the remark. The good lady was half inclined to be angry, but when she looked at Eleanor's smiling face, the anger evaporated, and she merely said : "You grow worse and worse every day, my dear daughter! I scarce know how to manage you."

"Manage me as you please, my dear, kind mother," said Eleanor, gaily, "only don't put the Protestant spectacles on me—let me look with the eyes that God gave me, undimmed by human prejudice. Now, Sir James Trelawney," she added, turning to him, "I know you are a seeker after truth, and that you are studying the character of

on people under a religious point of view—am I
not right ?"

" Perfectly so," said Trelawney, with a slight
bow.

" Then I will just ask you to accompany us in a
visit which we are about to make, and you will see
the Catholic religion in full operation."

" What, Eleanor !" cried her mother, " do you
mean to bring Sir James into Bernard O'Daly's ?"

" Even so, my dear mother."

"I shall be but too happy, Miss Ousely," said
the baronet, with even more than his usual suavity,
" to make any visit in such company."

" Nay, no compliments," said Eleanor, laughingly ; " bottle them up, and they will keep for those
who require them—we here are plain country folk,
you know. But, hush ! there is the house—you
see it is just on our way. Are you coming in,
mother ?"

" Yes, my dear, I believe I shall. Ben, pull up a
little—we want to stop at Bernard O'Daly's. You
can walk the horses up and down the road a little
way, till we come out."

Trelawney was instantly at her side to hand her
out of the carriage, while Eleanor stepped lightly
out, without waiting for assistance, and was the
first to enter the house.

Bernard met her at the door, his eyes red and

swollen. "How is Mrs. O'Daly?" said Eleanor, in a low voice.

"As bad as she could be, Miss Eleanor dear! oh! dear me, Mrs. Ousely! is this you, ma'am? why then, indeed, I didn't expect to see you here. Won't you sit down, ma'am? an' the young gentleman—please to take a seat, sir!" Having seen the visitors seated, the old man went to the room door, and made a sign to Kathleen to come out. The young woman was somewhat startled on seeing a strange gentleman with the ladies, but she quickly recovered her usual quiet composure.

"God bless you, Miss Eleanor dear! you didn't wait long to fulfil your promise."

"How is your mother now, Kathleen?" inquired Mrs. Ousely.

"Very low indeed, ma'am, thanks to you for asking; Father O'Driscoll is with her now—he gave her the rites of the Church this morning, and he had us all praying there in the room, when we heard the carriage stop. Wouldn't you wish to see my poor mother, ma'am? I know Miss Eleanor would!"

"And I, too, Kathleen," said Mrs. Ousely, "if our presence will not disturb her."

"Oh, no fear of that, ma'am—it's past that with her." Poor Kathleen's voice failed her, for just then there came a voice of wailing from the room. "It's little Eveleen, poor child!" mur

mured Kathleen, " God pity her !" The tone was
that of " God pity us all !"

" I'll jist go in and tell my mother that you're
here," said Kathleen. She went in, leaving her
father with the visitors, and in a few minutes
returned, making a sign for them to go in. " You
can stay here at the door of the room," said
Eleanor to Trelawney, " so that you may see and
hear what passes within. We shall not keep you
long."

" And here's a chair, sir," said Bernard, taking
another at a little distance. Mrs. O'Daly was
sitting up in the bed, supported by pillows, for her
disease was of an asthmatical kind ; her breathing
was hoarse and rapid, and her eyes wandered
restlessly around, as she gasped and struggled for
breath, in a manner pitiful to behold. Her face
was ghastly pale, and the nose was already pinched
and sharp, a sure harbinger of death. The priest
was seated in a chair beside the bed : Bridget was
on the opposite side, with one arm around her
mother's neck, while with the other she alternately
wiped the cold dew from her forehead, and fanned
her face with her handkerchief. None of the sons
were present, and Eleanor thought it strange that
they should be absent at such a time.

Father O'Driscoll bowed as the ladies entered,
and would have resigned his seat, but Eleanor, in
a low voice, begged him to remain where he was.

The sick woman looked round, and seeing Mrs. Ousely, she bent her head, but to Eleanor she reached her hand, and made an effort to say, "I'm glad an' thankful, Miss Eleanor dear! It's very good—of you, ma'am, to come to see—a poor creature—like me! Kathleen! bring chairs—for the ladies." A violent fit of coughing here set in, and while it lasted, the two girls held their mother up, then laid her back exhausted on the pillow.

"My dear Mrs. O'Daly!" said Mrs. Ousely, "I am very sorry to see you so poorly! We should have been to see you sooner, had we known any-thing of it."

"I'm thankful to you, ma'am, for all your good-ness—to me. You were ever an' always—kind an' thoughtful, an' if you had been to the fore, or Miss Eleanor either, a Monday mornin' last, the master would never have treated me as he did. But it was to be—ochone! I suppose I had it to go through."

"My dear child!" said Father O'Driscoll; "you had better say nothing about that. You are too weak in body and in mind to bear any excitement, and besides, you have promised to forgive and forget. Remember that, my child! remember the words of your daily petition: 'forgive us our trespasses, as we forgive them who trespass against us!'"

"I *do* remember it, your reverence, I do indeed,

an' though I said them words to the misthress an.
Miss Eleanor, I had no harm in them, Father
O'Driscoll—oh no, sir ! God knows I can say from
my heart out that I bear no ill-will to any one.
All that troubles me now is that I must leave
Bernard and the children !"

Here Eveleen sobbed aloud, and her sisters
could not restrain their tears The priest admon-
ished them in a whisper, not to disturb their
mother, and then turning again to her, he said :
" And why trouble yourself about that ? You
are going to a region of endless joy, where, after a
little while, you shall see all those you love again.
You have brought your children up in the fear and
love of God—they will work their way bravely
through the trials of this life, and then they shall
all go in turn to rejoin you in heaven. Till then
you will pray for them, and you can thus do more
for them than if you were with them here on earth.
Be not uneasy, then, about your family. Resign
them all into the hands of God, and beg of the
Blessed Virgin to be a mother to them when you
are gone."

Honora raised her hands and eyes to heaven,
and her lips moved in prayer, but no words came
forth. Gradually her face lost its sorrowful expression, and a look of benign tranquillity stole
over the shrunk and wasted features. Eleanor
and her mother feared she was dying, but Father

O'Driscoll assured them that she was not so near death as they might suppose. "She will hold out," said he, "in all probability, till the turn of the night."

"But where are the young men?" asked Eleanor of the priest.

"They are away working their day's work," said he with emotion, "digging out a ditch for Mr. Dixon."

"Is it possible, sir?"

"Ay! indeed, Miss Eleanor!" said poor Honora, who had heard what the priest said, though he had spoken almost in a whisper. "The poor boys took their spades in their hands yesterday mornin', an' went to ask work from Mr. Dixon. God help them, poor fellows! it's little we thought a year or two back, that they'd be workin' in a ditch shough for sixpence a day! Well! it's bet for us that we can't see what's before us—och! 't is, indeed!"

Father O'Driscoll again interposed with his consoling voice: "And don't you know it is for your sake they do it, my dear child? You have reason to be truly thankful that God has given you such children!"

"Ay! ay! sure I know it's to buy some little comfort for their poor sick mother that they took it upon themselves to go—och! God forgive me for this sinful pride—this foolish pride that strks

14

to me. O Lord! root it out of my heart, an'
give me the grace of true humility. Make me
thankful, O my God! for these little trials, for
ochone! but I wanted something to humble me!
Father O'Driscoll! with God's help you'll not
hear me grumblin' any more about our poverty—
I'll take up my cross now, late as it is, an' I'll
meet my Judge with it in my hand. There, Kathleen dear! lay me down. I'm weak, children,
weak, weak!" She closed her eyes, and lay a
few minutes motionless, but hearing the ladies
move, she opened her eyes and fixed them on
Eleanor. "Come here, Miss Eleanor!"

The young lady approached, and bent her head
to listen.

"Tell your father," said she slowly and with
difficulty, "tell him Honora O'Daly forgives him.
But tell him too, miss, that if he goes on as he's
doin', persecutin' the poor cratures for their religion, he'll bring down a curse on himself an' all
belongin' to him. Tell him that from a dyin'
woman. Bend down your head nearer, Miss
Eleanor:" she did so, while her tears fell fast on
the pale face of the dying woman. "I want to
leave you all the legacy I can—be a Catholic,
Miss Eleanor! if you want to save your soul.
If you do, I'll not bid you good bye for ever, we'll
meet again in heaven. If you don't, may the Lord
pity you!—you needn't blame Honora O'Daly!"

"I thank you sincerely," said Eleanor, with a blanched cheek and a tremulous voice. "I shall not forget your warning! Farewell, Mrs. O'Daly! I hope to see you to-morrow."

Honora smiled and shook her head. "If you come to-morrow, Miss Eleanor, it's these," pointing to her husband and her weeping daughters, "it's these that will want comfort, not me. I'll be gone on my long journey before then. Do all you can for Bernard and the children, Miss Eleanor, they'll need friends, God help them! till Cormac can send them relief!"

Mrs. Ousely then shook hands with Honora, and told her she would send down some things for her use as soon as she reached home. "Thank you, ma'am," said Honora faintly, "I don't think you need take the trouble—I don't want much now, my eatin' an' drinkin's near over! God be with you, ma'am, you have the good wish of the poor every day you rise, but they'd think far more about you than they do, if you'd let them alone about religion, an' do as Miss Eleanor does."

"Well! well!" said Mrs. Ousely, smiling pleasantly, "perhaps I'll behave better for the time to come. I see you Catholics are very different people from what I thought. Farewell, Honora! Come, my daughter, it is wearing late!"

Meanwhile, Father O'Driscoll had joined Bernard and Sir James, and had entered into

conversation with the latter, chiefly on the solemn
scene before them. When Mrs. Ousely and
Eleanor came out, the latter introduced them to
each other, for Father O'Driscoll and she were
old acquaintances, having often met on similar
occasions, at the bed-side of the sick and dying.

"I regret, sir," said Trelawney, "that we must
part so soon. I should have wished to cultivate
your acquaintance a little more, had time permitted."

"If you are not engaged to-morrow, come and
dine with me then!" said the priest with a smile,
as he shook hands with Trelawney. "Any one
can show you my humble domicile."

"I shall certainly avail myself of your kind and
welcome invitation, sir," said the baronet with a
graceful bow. Having seen the ladies seated in
their carriage, he would have wished them good
morning, but Mrs. Ousely insisted that he should
see them home, "and unless you are otherwise
engaged," said she, "you must stay and partake of
our family dinner. Mr. Ousely will be more than
pleased to have your company for the evening."

"So be it, then," replied Trelawney with a
smile. "Persuasion is easy, you know, where
inclination leads th way."

CHAPTER VIII.

"The keen is loud, it comes again,
  And rises sad from the funeral train ;
  As in sorrow it winds along the plain,
    By the bonnie green woods of Killeevy

"Death is not always an evil."

THE carriage had scarcely left the door, when
Granny Mulligan stept out from behind the barn
and tramped into the house, bending under the
weight of a well filled bag. Seeing no one in the
kitchen, she called out: "Come here some of ye,
children, an' take this bag off my back."

Kathleen hastened out from the room, and told
Granny in a whisper, as she lifted the bag from her
back, how Mrs. Ousely and Miss Eleanor, and the
English gentleman, from Clareview, had been to
see her mother.

"I know they were, Kathleen," said the beggar-
woman quietly, "my back an' shoulders can tell all
about it, for when I got to the end o' the barn, an'
seen the coach at the door. I didn't want the quality
to see me comin' in here with my bag full, so I just
waited there till they'd be gone, an' a good stay
they made of it. But how is your mother—did
she get e'er a turn since mornin'?"

14*

"No, granny, she's much about the same way - only may be a little weaker—but sure Miss Eleanor is to send down some nice things for her as soon as she gets home. I don't know what I'll do for the boys' dinner—I haven't more than a dozen of potatoes."

"An' where's your eyes, Kathleen, that you don't see the bag beyant—isn't there enough there. for two or three dinners? Go off now, an' wash the praties an' I'll put on the pot, an' we'll have the dinner in a jiffy. What are you gapin' at me that way for, you foolish *colleen*—go an' do what I bid you."

"Well, but, granny—"

"Don't be botherin' me now with your talk, Kathleen," said the old woman, sharply, "I must go an' see what way your mother is in, an' mind you have the dinner ready soon, for I'm goin' up to Clareview by-an'-bye, and I'll take it to the boys."

"No, no, granny," said Kathleen, "I'll send Eveleen."

"No, nor the sorra step you'll send her—do you think I'd let Eveleen, or any of you, go on that errand, an' me here? *Musha*, but you're the quare Kathleen, to think o' the like. No, no," muttered the kind-hearted old woman, as she hung up her cloak, and turned into the sick room, "no, no—it's bad enough as it is."

"Well! God reward you, granny-—that's th
best I can wish you!"

"Never fear but he will," said Granny, at the
door of the room, "I'm not much afeard about
that."

"What is that you say, granny?" asked Father
O'Driscoll, who was still sitting at the bedside.

"It's talkin' to Kathleen I am, your reverence,
about a little matter that's atween ourselves. How
do you feel now, *astore?*" putting her hand on the
sick woman's head.

"Neither better nor worse, granny dear," replied
Honora, in a low, husky voice.

Granny Mulligan said nothing, but she looked
significantly at Father O'Driscoll, who shook his
head, and made a sign for her not to speak much.
Bridget and Eveleen looked into granny's face to
see what she thought, and their tears began to flow
again, when they saw the mournful expression so
visible on every feature—they knew that she had
no hope. Bernard, who was sitting sad and silent
in one corner of the room, brought over a chair to
the bed-side, and resumed his place without saying
a word. In a few minutes Father O'Driscoll arose,
and saying that he would be back in two or three
hours, was about to leave the room, when Honora
stretched out her emaciated hand, and murmured
"Well! God's blessin' and mine be with you,
Father O'Driscoll!—you've done your own share

for me an' mine, any how. I have one favor to ask of you before you go, for fear I'd be gone before you come back."

" What is it, Honora ?" inquired the priest, bending his head to catch her faint accents.

" Won't you say a mass or two for me, as soon as you can—I've no money to offer you, but I know that will be no hindrance to your charity."

" Make your mind easy on that head, my poor child !" said the priest, with emotion; " I'll not forget you, be assured of it. But I hope to see you again before——"

" Before I go—God grant that you may, your reverence ! I'd like to have you near me at my off-goin', but if anything keeps you away, why—the will of God be done !" She closed her eyes, in silent meditation, and the priest moved quietly away, followed to the door by Bernard, who stopped him on the outer threshold, to ask how long poor Honora was likely to hold out.

" She may last till midnight," said Father O'Driscoll, as he shook Bernard's hard, and squeezed it hard, " but it is much more likely that she will drop off about night-fall. God comfort you, Bernard !" The old man raised his tearful eyes to heaven, but he could not speak, his heart was too full.

Before Bernard had returned to the sick room a messenger arrived from the Hall, with a well filled basket, containing wine, tea, sugar, some loaves of bread, and several other little matters useful for the sick. Kathleen was called to put away the things, and all the time she was thus employed, her heart was raised in thanksgiving to God, and in earnest supplication for the spiritual and temporal welfare of the generous donors. Never did the thought once cross her mind that the Ousely family owed hers more than this—a thousand times more. When Kathleen returned the basket to the servant, he said that Miss Ousely sent her compliments to know how Mrs. O'Daly was.

"Miss Ousely is very kind," said Kathleen, " but no kinder than I would expect her to be. Tell her that my poor mother is just the same way, and that we don't expect her to get over this evening. Give her and Mrs. Ousely our best thanks!"

Towards evening the young men came in from their work—their first day's work for the stranger, and the first question was, "How is mother?" Bridget, who had resigned her place in the sick room to her elder sister, was now engaged in preparing the supper, and she answered only by a sorrowful shake of the head, and a fresh burst of tears.

"So she's no better, Bridget?" said Cormac in a whisper.

" Worse, if anything, Cormac dear!

" Who's in the room with her—is there any stranger ?"

" No, only Phil Maguire and Nanny."

The brothers waited to hear no more, but hastened to their mother's bed-side. She lay with her eyes closed, and her cold clammy hands extended over the bed-clothes, without even the slightest motion. Seeing that the young men started on beholding her, Nanny Maguire told Cormac in a whisper that she was not yet dead. Cormac ejaculated his fervent thanks to God, and though he spoke below his breath, yet his voice reached his mother's ear. She opened her eyes, now dim and glassy, and tried to reach out her hand, but could not, and it fell powerless on the bed. Cormac took the hand and squeezed it between his own, as though he would warm it.

" It's no use, Cormac *aroon !*" she whispered, with a faint smile, " it's the coldness of death that's in it. Thank God—oh! thank God that you came in time. Are the boys there? Owen and Daniel, where are you, children ?"

" Here, mother darling, here we are !" and both burst into tears as they pressed up close to the bed. " Mother! mother! sure you'll not leave us ?" sobbed Owen, " Oh! what would we do without you, at all ?"

" God will do for you, my poor fellow, an' sure

I leave you in good hands, the Blessed Virgin will
be your mother. Who's that at the foot of the
bed? ah! that's Eveleen—come here, Eveleen,
my little one, my helpless one!" The little girl
laid her tearful face close by her mother's on the
bed. "Stay there, Eveleen, don't leave me any
more—you'll not have to wait long, dear! Oh!
but don't be cryin' that way, you'd only disturb
me when I ought to be quiet."

"Mother dear," said Kathleen, "don't talk so
much, it will do you harm." Nanny too begged
of her to keep still, but she only smiled and talked
on, whenever she could get out a few words.

"Where's Bridget? I don't see her." Bridget
came in, and then the dying mother cast a glance
around, resting a moment on every dear face.
When she came to Bernard, she made a desperate
effort, and succeeded in reaching out her hand.
"Poor Bernard!" she muttered, "you may well
cry, you're losin' one that loved you better than all
the world. But then, sure we're not partin' forever, we'll meet again, Bernard, never to part any
more. Take good care o' the children, Bernard
dear, an' see that none o' them falls away
from the service o' God—pray for them, aroon!
while you're left here behind me, an' I'll pray for
them when I get to heaven, which I will one day
or another, with God's help."

"Onny dear!" said Bernard, "I know all you

want to say—we all know it and with the assistance of God's grace, we'll do as you wish. Don't
be wearin' yourself away talkin'—don't, *alanna
machree!*"

"I b'lieve—I can't say much more—at any
rate!—Boys! there's the sound of a horse's feet.
Run out an' see—maybe it's Father O'Driscoll!"

"It is, indeed, mother," said Cormac, as he
returned with the priest.

"Och! thanks be to God!" said Honora fervently. "I'm a'most over, your reverence, God
sent you just in time. Where are you all, or did
you put out the light? I can't see." The priest, by
a wave of his hand, restrained the general outburst
of sorrow which these last ominous words called
forth, and he calmly commanded all to kneel, while
he read the prayers for the dying.

"Put the beads in my hand," said Honora,
"there—Kathleen dear!—I can't see you, but I
know it's you—put them that way—on my breast
—ah! Granny Mulligan! I hear you—pray for
me when I'm gone—an' Phil an' Nanny!"

She spoke so low, that Kathleen had to bend
down over her to catch her words.

"Wern't they in the room, lear?"

"Yes, mother, they're all here—praying for
you."

"That's right, Kathleen!—now, children, farewell!—Bernard! it's the poor, lonely Bernard

you'll be now!—God bless you all—God bless
you!—Cormac!—Eveleen!—God and the Blessed
Virgin be your guide!"

These were her last words—she never spoke
again. The prayers were read—the responses
went up in fervent unison from every heart—
Honora's lips were seen to move, and a smile came
over her wasted features, but neither foot nor hand
moved. At last the priest repeated the final act;
" Depart, Christian soul! go forth from this
world, &c." When it was ended, the smile was
still on Honora's face, and the hands were clasped
over the beads and crucifix, but the lips moved no
more: the soul was already before the Judgment
seat. The priest bent down over the dead, to sa-
tisfy himself that all was over; then raising his
hands and eyes to heaven, he said in a tremulous
voice: " May the Lord have mercy on you,
Honora O'Daly!"

This was the signal for the long-repressed sorrow
to burst forth. The girls threw themselves on
their knees beside the bed, and buried their faces
in the clothes, till they satisfied themselves—they
wept for some time, unnoticed and unrestrained,
for all present were more or less sharers in their
sorrow. Bernard sat down in a corner, and cover
ed his face with his hands, nor moved till the voice
of Father O'Driscoll drew him from his lethargy
of grief. Taking him by the hand he led him out

Into the kitchen, saying : " Come here, Bernard, I want to speak to you."

The old man followed with the docility of a little child, but as he passed the bed, he cast a glance at the still pale face of his dead wife, and muttering, " Poor Honora!—och! och!—is that the way with you at last ?"   He said no more, but went at the priest's bidding, and sat down by the fire-side in the kitchen.   Father O'Driscoll then reminded him that he had but little cause to mourn Honora's death, at least as far as herself was concerned, " for," said he, " the exchange is a happy one for her."

" Och! I know that, Father O'Driscoll !—I know that well ; but still—God help us—we can't help grievin' for our own loss. I know she's better off, your reverence, but then she's gone from us."   He looked over at the high-backed chair—now empty —and he could say no more.   The priest sat calmly by till the old man had " cried his fill," as he said himself, and then he talked with him of the exceeding great happiness of the " just made per- fect," and of the reward reserved for those who suffer all things for God's sake, until the bereaved husband began almost to rejoice that his poor, broken-hearted Honora had at length found rest and peace.   This was the frame of mind to which Father O'Driscoll had sought to bring him, and so, having spoken a few words of consolation to each of the bereaved children, he mounted his horse, and took his way home, with a promise to return in the morning and say mass. Phil Maguire and Cormac walked with him, on foot, part of the way, talking of the many virtues of the dead. The priest asked Phil whether he and Nanny proposed staying at the wake.

"Oh, blood alive! to be sure we do, your reverence. Nanny's goin' to help granny Mulligan now to lay poor Honora out, an' after that we'll stay all night. Oh! that's the least we may do, Father O'Driscoll, an' sorry, sorry we are to have the occasion."

"Thank you, Phil," said Cormac, his voice quivering with emotion, "I trust in God it will be long before any of us will be called upon to do a like kind office for you or yours."

"Well, now," said Father O'Driscoll, "I think you had better return home—I can go on alone. and your sisters will be looking for you, Cormac."

"We'll be biddin' your reverence good night, then," said Phil, "wishing you safe home. Why. who's this comin' along at sich a rate?"

The night was not so dark but that objects were distinctly visible, and a horseman was now seen dashing up the road at full speed. He was passing by, without noticing any one, when the priest called out: "Why, Tim Flanagan, is that you—where are you going in such a hurry?"

The man instantly stopped, and putting his hand to his hat, or rather cap, exclaimed: "Ah! then, Father O'Driscoll, sure enough it was God that sent you here. I was at the house, your reverence, lookin' for you, an' Nancy Breen sent me up to Bernard O'Daly's after you."

"Why, what's the matter, Tim? Is there anything wrong?"

"'Dt ed an' there is, your reverence—that unfortunate brother o' mine was taken very bad with a colic last night, an' there's no life expected for him. He's cryin' out for you, now, the poor unlucky scape-grace, an' I hope you'll forgive an' forget Father O'Driscoll, an' come away to him at once. There's no time to be lost, for he's a'most gone as it is!"

"Ah! the blackguard!" said Phil Maguire, "how has he the face to send for the priest, afther doin' what he did—it's a mortal sin not to let him die like a dog. Why doesn't he send for the ministher, Tim?"

"Oh! you're there, Phil Maguire, myself didn't look who was in it, my mind's in sich a state about poor Jack. Don't be too hard on him, Phil dear, for though I never exchanged words with him since misfortune an' *the soup* made him join the Jumpers, still I can't bear to hear him run down, now that he's sorry for what he done, an' wants to get back into the Church before he dies."

"Smd I thanks to him for that," said Phil gruffly, "he staid away as long as he could—he's no sich fool as to lose the last chance of savin' his sowl, but God grant him the grace of true repentance any how! Myself hopes that he'll not die before the priest gets there, though we all know that he deserves to be taken short; howanever, God's merciful!"

"Come, come, Tim!" said the priest, who kept moving on during this brief colloquy; "let us pull out—thank God! we have not far to go, and we *may* still be in time. Ride now, Tim, for life and death—there's a precious soul at stake!"

Cormac and Phil stood listening on the road till the clatter of the horses' feet died away on the still night-air, and as they turned to retrace their steps, the young man said with a heavy sigh, "My poor, poor mother! how often have I heard her prophecy what has this night come to pass! She used to say, when she'd hear of Jack Flanagan's ridiculing the priests, that whoever lived to see him in his last hour, he'd be calling for the priest, and calling in vain."

"Ay! an' it's ten chances to one if he don't die with the word in his mouth—there's something tellin' me, that Father O'Driscoll won't get there in time! Oh, Cormac dear, but it's a foolish thing to depend on a death-bed conversion! See what a difference there is between the death of that un-

lucky cratcre, if so be that he does die, which I
suppose he will, an' your mother, the Lord rest
her sowl in glory !"

"Yes! Phil, that is our only consolation—my
dear mother's death was just an instance that ' as we
live, so we die'—it was a fitting end for a life of
simple piety and unpretending virtue !"   Beautiful
and touching was that heartfelt tribute to a mother's
memory, and so thought Phil Maguire, but he said
nothing.   His sympathy was too deep for words.
Just then, too, they reached the door, and were
met by Granny Mulligan, who told them, in her
bustling way, not to go ' near the room' for another
while, "for," said she, " Nanny an' myself's doin'
something in there that we don't want to be dis-
turbed at.   The people's beginnin' to come in for
the wake, Cormac *aroon!* so jist go, you an' Phil,
an' get the pipes an' tobacco.   Them things that
came from the Hall are in good stead now !
Kathleen an' Bridget's goin' to make tay for the
women by an bye !"

"Well! go in you, Phil, and send Daniel or
Owen here," said Cormac, " we must go into town."

"An' won't I do as well as either Daniel or
Owen ?" demanded Phil, testily.   " Come on, now,
Cormac, an' don't be standin' there dilly-dally !"

Phil knew well what he was about.   He in
tended to make the necessary purchases himself,
and would have been just as well pleased could

he have gone alone, but, knowing that he could not
well get rid of Cormac, he must only manage it
the best way he could.

A busy woman was granny Mulligan all that
long night. Every member of the family being
too much engrossed with their own heavy sorrow,
to pay proper attention to the neighbors and friends
who thronged in to the wake, granny took it upon
herself to receive everybody, to show everybody
to their proper places, and to see that everybody
had what was needful and fitting for them. The
old people she ushered into the room, where
Bernard sat in speechless woe near the foot of the
bed, whereon the corpse was laid out. The young
people were all placed in the kitchen and in the
young men's bedroom, cleared out for the occasion. Nanny Maguire had her own share of the
duty to perform, and she bustled about the fire-
side, superintending and helping on the preparation
of "the tay" aforesaid. A trifling dispute sometimes arose between her and Kathleen regarding
the quantity of tea to be put down, or some such
thing, Nanny still insisting that there was "no use
in puttin' down so much"—that "enough was as
good as a feast," and that "wilful waste made
woeful want." Kathleen would smile a mournful
smile, and say, "Let us have enough for this one
night, Nanny dear! My poor mother never

stinted any one in eating or drinking. when she
had it to give, and now when God and Miss Eleanor sent us plenty, let us give as freely as we got."

" Well! well! have it your own way, Kathleen,
honey! but I declare to my sins, it goes to my
heart to see sich waste, and in times like these,
too."

When Cormac and Phil returned, the pipes and
tobacco were laid on the tables in each room, and
the house was soon reeking with the smoke and
smell of tobacco. After a little there came a
knock at the door for the twentieth time, and when
it was opened, who, of all the world, should be
there but Andrew McGilligan, his tracts, as usual,
under his arm. One looked at another, but no one
spoke, and Andrew looked round in vain for a seat
—he might have looked long, for the spare seats
were all slily shoved into corners, out of sight, as
soon as his doleful countenance had appeared within the door. It was granny Mulligan who first
spoke, in right of her self-appointed office of mistress of the ceremonies.

" Well! good man—what's your business here ?"

" I have just heard that there is a wake in this
house to-night, and knowing the profane sports
usually practised on these occasions, amongst Romish people, I was moved to come and provide
for the numerous company here assembled an en

tainment far more profitable and more becoming
for the house of death."

"An' what sort of entertainment are you goin'
to give us, *zgra?*" said granny, putting her arms
a-kimbo, and planting herself firmly on her feet.
At the same time she winked at the amused and
expectant listeners.

"I have brought some excellent pamphlets, from
which I can choose some interesting narrative to
read for these good people's entertainment. Can
you accommodate me with a seat, my good old
lady ?"

"Ho! ho!" cried granny, "I'm a good old
lady, now, am I?—ah! then, Andy, *acushla!* is
that the way with you now?—don't you mind the
other day when you gave me all the abuse in the
world, bekase I asked a charity at Jack Flanagan's
—I was an old Popish vagrant, then, but now I'm
'a good old lady'—lady, *inagh!*—don't I look like
a lady—eh! boys an' girls?—don't you think Andy
can lay on the blarney thick? Now I'll just tell
you what it is, Andy McGilligan; it's the best of
your play to make your escape as fast as you can
—if the people o' the house sees you, I can tell
you you'll not be thankful to yourself for your
visit."

"But surely you will not deny me a seat?"

"The seats are scarce the night," returned

granny, shortly; "we have none to spare. What are you drawin' s: near that room for?"

"My dear woman! I see some of the family in that room, and would wish to administer comfort unto them." He was still making for the room door, whereupon granny placed herself directly in his way, and waxing warm upon it, shook her fist in his face.

"I tell you now, once for all, Andy McGilligan! that you shan't set your foot inside o' that room. Why, man, Honora O'Daly couldn't rest in pace if she knew that there was one o' your tribe near her. Away out o' this with you now—tracts an' all! or that I mayn't do harm, but I'll try the strenth o' my arm on you—ould as I am, I think God would give me strength enough to bate a Bible-reader."

"What is this, granny, what is this?" said Cormac, coming out from the inner room.

"There now," cried granny, in a high state of excitement, "you wouldn't go till you brought Cormac out. See there, Cormac *aroon!*—there's Andy McGilligan forcin' his way into the room— he wants to comfort you with some tracts he has here."

The young man fixed a look of scathing scorn on the luckless Bible-reader, but he merely said, in a thrilling whisper: "Man! man! will not even the presence of death screen us from your perse-

eution ?" and without waiting to hear a word of
Andrew's attempted justification, he quietly led him
to the door, and was preparing to shut it after him,
when a lad who had entered but a few minutes
before, called out from within :

"Arrah ! Andy, did you hear the news ?"

"No !" returned Andrew from without : "what
news ?"

"Jack Flanagan's dead, an' he died jist as he
ught to do, in black despair. He was shoutin'
for the priest from ever he found death upon him,
an' his brother Tim went off post-haste for Father
O'Driscoll, but when they came, the poor divil was
speechless, an' workin' for death, an' so the priest
could do nothing but kneel down an' say a prayer
for him. He died without being able to say a
word, an' they say it was terrifyin' to see him."

"I do not believe you, my good young man. I
cannot and will not believe you."

"You may do for that as you like, my good
young man—as you say yourself—but what I tell
you's true enough, an' what's more, it will be your
own story some of these fine days, unless you alter
your ways."

A low titter passed around amongst the young
people, and Cormac hastily closed the door against
the crest-fallen agent of " The Protestant Missions,"
then beckoning Phil out of the room, he told aim
in a whisper of Flanagan's unhappy death.

"Poor wretch! poor wretch!" ejaculated Phil, his kind heart touched with sorrow for the man's miserable end; "may the Lord forgive him his sins—I hope it's no harm to pray for him!"

"An' I wish it may be any use either, Phil!" chimed in Nanny from behind; "the best thing we can do is to take warning by his example, an' pray for the grace of a happy death."

Early next morning Eleanor Ousely walked down to Bernard O'Daly's, anxious to know whether Honora O'Daly was living or dead. When she came to the door and saw the kitchen full of people, she knew at once that the poor weary spirit was released from bondage, and there was a sort of melancholy pleasure in the thought. Her appearance was evidently unexpected, for the people, old and young, started and stared, but all stood up and bowed low, and smiled in answer to the young lady's graceful salutation, and many a fervent "God bless you!" arose from heart and lip to heaven, for Eleanor Ousely was the protectress of her father's poor tenantry, their advocate, and their benefactress. Bernard was not slow in making his appearance, and Eleanor, taking his hand kindly, looked sorrowfully towards the shrouded corpse, visible through the open door.

"Yes, she's there, Miss Eleanor," said Bernard, following the direction of her eyes, "she can't welcome you now. She's gone from me at last.

after our long partnership. Well! God's will be done, anyhow! Won't you come in an' sit down, Miss Eleanor dear? The girls are in here, except poor little Eveleen, that we sent to bed about an hour ago."

The two sisters were sitting sad and sorrowful beside the bed, and on seeing Eleanor they both burst into tears, remembering how much their poor mother had loved her, and they could neither of them speak for some time. There was only Phil Maguire and one or two others in the room, for Nanny and granny Mulligan had been persuaded to lie down for a few hours, after the fatigue of the night.

About nine o'clock Father O'Driscoll arrived, and a temporary altar was quickly prepared in the room with the corpse. While the preparations were going forward, the priest approached Eleanor, and said, in a low voice: " Miss Ousely, perhaps you would rather not be present during Mass—if so, you had better withdraw till it is over—it will not be more than half an hour or so."

" You are very kind, Mr. O'Driscoll, to think of me, but I am not afraid of ' Popish rites,' as Andrew McGilligan would say. I have no objection to worship God with Catholics."

" Then *you* do not consider us idolaters, Miss Ousely?" said Father O'Driscoll with a smile.

" Not exactly, sir," and Eleanor smiled too.
16

" It is very doubtful, indeed, whether any one does really *consider* you as such, but it is very certain that there are many who *call* you so, for reasons well known to themselves, and others too."

" I am rejoiced to hear you speak so, Miss Ouse-ly," said Father O'Driscoll. " Brought up as you have been, one could scarcely expect you to do Catholics so much justice."

" Oh ! my training has not been quite so bad as you would imagine, my dear sir !" said Eleanor, warmly. " I was so fortunate as to have a precep-tress whose mind rose far above vulgar prejudice, and who was impressed with profound veneration for the Catholic Church. With her I studied ecclesiastical, as well as sacred and profane history, and I am, therefore, well aware of the claims which your Church has upon our respect, I will not say submission."

" Having gone so far, then, my dear young lady, how *can* you remain as you are, cut off from that Church whose true character you seem to under-stand ?"

" Nay, that is another question," replied Eleanor quickly ; " I trust I am *not* cut off from the Church —I belong to an arm of the universal Church—I mean the Church of England."

Father O'Driscoll smiled, and shook his head. " It won't do, my dear Miss Ousely, it won't do ! When a member is dissevered from the body it

cannot exist alone—the principle of life remains
with the body, as you cannot but admit. The
Church of England was three hundred years ago
lopped off from the body of the Church, how then
is she a part of it? Take care, my dear young
lady, take care how you tamper with a matter
which concerns your immortal soul!"

Nanny Maguire now came in, with several others of the neighbors, and the priest, seeing that all
was in readiness, prepared to say his mass, served
by Daniel and Owen. During the time of mass,
Eleanor knelt with the others, and it seemed to
her that she had never prayed with so much fervor. There was inspiration in the rapt devotion
of the simple cottagers who knelt around, and in
the solemn presence of the dead, for whose
eternal repose their prayers were offered up. The
last words of the priest were uppermost in Eleanor's mind, and she could not help asking herself
the startling question, "What must I do to be
saved?" She looked at the priest, offering up an
atoning sacrifice—a renewal of that of Calvary—
"at least that is their belief," said she to herself—
then at the numerous relatives and friends, praying
with heartfelt devotion for the departed—then at
the lifeless clay that was soon to be consigned to its
parent earth, and she said within herself, in the
words of Judas Macabeus, "It is a good and

wholesome thought to pray for the dead, that they
may be loosed from their sins."

Mass being over, Father O'Driscoll called for
his horse. Bernard would fain have kept him for
breakfast, and Kathleen, too, tried her persuasive
powers, but all in vain—stay he would not.

"No, no, Kathleen, don't ask me—that's a good
girl—I must hurry home. Don't blame me, Bernard," he said in a low voice, but still Eleanor
heard what he said; "You know I'd rather eat
potatoes and salt with you, than partake of the
daintiest fare with others. but then you have
enough for breakfast without me, and times are
not as they used to be."

"Well now, your reverence, that bates Banagher," said Bernard, "but sure it's just like you
not to let your right hand know what your left
gives. Didn't Nancy Breen come over this mornin'
early with as much as would make three breakfasts—
so the girls and Nanny Maguire says—maybe we
have just as much here as *you* have at home."

"Well! well! never mind, Bernard—excuse me
for this time. I'll come over to-morrow and say
mass again before the funeral goes out—I'll
breakfast with you then. Good morning, Miss
Ousely! I hope you will think of what I told
you this morning." Eleanor bowed assent, and
Father O'Driscoll retired, after speaking a few
words with Phil Maguire. Eleanor remained

but a few minutes after the priest, and when she was going, she tried to prevail on E-releen to go home with her until after the funeral. But Eveleen would not hear of such a thing, even though Kathleen and Bridget advised her to go. "How would you like to go yourselves, either of you?" said she, sobbing, "and to leave my poor mother that will not be long with us now. Indeed, Miss Eleanor! I wonder you'd ask me to do such a thing." So saying, she escaped into a corner near the bed, and would not hear another word.

Next morning, when the appointed time was come, every relative and friend of the family having kissed the cold, pale lips of the dead, and bid her a long farewell, the lid was screwed down on the coffin, and the corpse of poor Honora O'Daly was taken from the house whose mistress she had been for well nigh thirty years. As the distance to the grave-yard was not more than a mile, the coffin was borne on men's shoulders, parties of four being appointed to relieve each other. Cormac and Daniel, with two of their cousins, took the first turn, and Phil Maguire insisted on being one of the second bearers. Bernard, with his son Owen and his three daughters, walked after the coffin, and behind them followed granny Mulligan, Nanny Maguire, and our two acquaintances, Judy and Nelly, whom we saw in the chapel-yard on the day when Katty Boyce told her story to the priest.

These four old women were the *keeners*, and sang
at intervals the mournful wail known as the Irish
funeral cry, now no longer heard save in the more
remote and mountainous regions, where the primi-
tive habits still prevail. After them came a
multitude of men and women attired in their best,
though that best was bad enough with many of
them. The funeral was one of the largest that had
been seen about Killany for many a day, and it
was a remarkable fact that, notwithstanding the
reduced circumstances of the O'Daly family, there
were several gigs and jaunting cars from the village
and the adjoining country, showing how much and
how far the family was respected.

Any one who has once heard the *ullelula*, the
Irish dirge—can never forget the unearthly wild-
ness, the mournful tenderness of the strain, and on
that day it startled the echoes of the hills, and
died away in faint cadence along the far-off shores
of Corrib; for many of the aged women who
followed in the funeral train had been playmates
and schoolmates of Honora, had known her from
youth to age, had received abundantly of the fruits
of her prosperity, and had borne a sympathizing
share in the reverses of her later years; they ever
and anon swelled the doleful chorus with the voice of
their deep, deep sorrow. At every cross-
road and at almost every house the funeral was
increased, and each, as they fell into the ranks of

he procession, murmured, "God rest your sowl
Mrs. O'Daly!" or some such fervent ejaculation.
Death is not in Ireland the cold, dull, dreary thing
that it elsewhere is; the warm, genial sympathy
of the Celtic, the Catholic heart, is a soothing balm
to the mourner's troubled soul, extracting the sting
from affliction, and depriving death of half its bit.
terness. Byron felt the truth of this when he
sang that well-known stanza :

' ——I had envied thy sons and their shore,
    Though their virtues were hunted, their liberties fled,
There was something so warm and sublime in the core
    Of an Irishman's heart, that I envied—thy dead."

Somewhat similar were the reflections of Eleanor and Sir James Trelawney, who, with Mrs.
Ously, were present in the chapel when the
corpse was brought in and laid before the altar,
while mass was said. It was the first time that
Trelawney had seen an Irish funeral, and the scene
made an impression on his mind that time could
never efface. He then beheld the Irish people under an aspect new to him—the deep-seated piety
and the exquisite tendernes, both characteristic of
the peasantry of Ireland, were there distinctly
visible, and from that day forward Trelawney
entertained a profound respect for that down-trod-
den yet most interesting people.

Mass being over, Father O'Driscoll threw off
his robes, except the alb and stole, and proceeded

to bless the grave wherein Honora O'Daly was to await the Resurrection. The ceremony was brief, and the coffin was soon lowered into 'the narrow house,' amid the sobs and lamentations of friends and relatives. Bernard threw in the first shovelful f earth, then the sons in succession, and in half an hour the green sod of the churchyard was smoothed over the grave, and the mourners having knelt a few minutes in prayer, took an unwilling leave of 'the loved' (but not 'the lost'), and 'the lone place of tombs' was left to its wonted stillness.

CHAPTER IX.

"They did not know how fate can burn
In hearts once changed from soft o stern,
Nor all the false and fatal zeal,
The convert of revenge can feel."

                              BYRON's *Siege of Corinth*

" The hand that oped spontaneous to relief,
The heart whose impulse staved not for the mind
To freeze to doubt what charity enjoined."

Sir James Trelawney accompanied the ladies home after their visit to the school, and was warmly welcomed by Mr. Ousely, who begged leave to introduce to the baronet a friend of his, the Rev. Mr. O'Hagarty, "formerly a priest of the Church of Rome," said Ousely, "but now a minister of the Church of England, and curate of this parish. And a cursed shame it is to have him a curate— which I call being put on dog's allowance. Sir James! you'll find the reverend gentleman a confounded gay fellow, I promise you. Walk in, Sir James, you're welcome to Ousely Hall, where I hope you'll consider yourself at home!"

Trelawney bowed his thanks, and having duly greeted the reverend gentleman, they both followed Ousely into a front parlor. Conversation did not flow very freely, for somehow O'Hagarty and the baronet were neither of them willing to talk a great deal, and though Ousely himself started a number of topics, they were, none of them, kep up for any length of time. At length Ousely chanced to ask whether Sir James had been to Jenkinson's school with the ladies. The baronet answered in the affirmative.

"Well, and how do you like it?—You must have been rejoiced to see such a number of Papist brats under good Protestant training! a d——d fine sight, is it not?"

"Pardon me, Mr. Ousely! I am not quite of your opinion regarding those schools. '

"The d—l you're not?" cried Ousely, while the ex-priest opened his sleepy-looking gray eyes as wide as they could stare. "And pray what is your opinion of the schools—I presume you have formed one?"

"Most assuredly I have!" replied Trelawney. "In the first place, I see no reasor why Roman Catholic children should be taken from their own rightful teachers, and subjected to Protestant training, as you say. What is the object of this, or on what principle of right can you justify it?"

"Justify it—justify it? Why, simply because it is always just and lawful to diffuse the ennobling spirit of Protestantism—"

"Yes," added O'Hagarty, "and to emancipate the mind from the slavish yoke of Popery. This,

sir, is or ought to be the grand object of all true
Protestants !"

"You say so, sir !" said Trelawney calmly.

"Yes, I say it, and I maintain it !"

"Well ! but the principle, sir—on what principle can you do this thing ?"

"Oh! as to the principle !" exclaimed Ousely,
snapping his fingers, "I don't care *that* for
principles. Protestantism *must* be spread, do you
see, by one means or the other, and I never trouble
myself with any scruples as to how it is to be done.
In religion, as in war, every stratagem is fair, so
long as it tends to promote the ultimate object.
Hang it, that's my notion. Let that pass, now, for
you'll get no more out of me. I hope you can
find no fault with the system of teaching there ?"

"No, no," said O'Hagarty, opening his box and
taking a huge pinch of snuff; "I am sure the gentleman must approve of the system ; Mr. Jenkinson is an excellent teacher—a capital teacher all
out."

"I am sorry to differ from you once again,
gentlemen," said Trelawney, "but as you have asked
my opinion, I must tell you candidly that the only
thing systematic in Mr. Jenkinson's teaching is his
constant abuse of Popery. Now it strikes me that
that is a very poor substitute for useful knowledge,
the knowledge of God and of our own dependence
on Him, with the various obligations which bind

as to Him and to society.   Such billingsgate abuse
of the Catholic Church might and *may* do if the
boys and girls are all intended for taking a part in
the fantastic exhibitions of Exeter Hall, but other
wise it is good for nothing."

"Why, deuce take me, Sir James!" cried
Ousely, with a horse laugh, "but I think you're
half a Papist yourself!"

"The gentleman is certainly no warm supporter
of Protestantism—that's plain enough, I think,"
observed the Rev. Mr. O'Hagarty, taking another
pinch of snuff, and then handing the box to Sir
James.

"Thank you," said the baronet with a slight
bow, 'I never take snuff.   But you are quite
mistaken, gentlemen, in supposing me to favor the
Catholics.   On the contrary, I was, until very recently, an energetic opponent of theirs."

"Oh ho! I see," said O'Hagarty, with emphasis,
"until very recently!—that's as much as to say
that you're not so still—eh?"

"I am not aware, sir," said Trelawney, haughtily,
"that you have any right to question me as to
what change my opinions may have undergone.
Mr. O'Hagarty probably supposes that the Oxford
graduates are all on an inclined plane, but I am not
of Oxford," he added, with a good humored smile.
"Cambridge is my *Alma mater*.   Excuse me,
gentlemen. here are the ladies!"   He arose and

went to join them at a distant window, where
Eleanor was pointing out to her mother the beautiful tints of the autumnal foliage in the woods
around.

"I come to take shelter with you, ladies!" said
Trelawney. "Those gentlemen are bent on po-
lemics, and I have left them to talk the subject out
between themselves."

"You were in warm quarters there," observed
Eleanor archly, as she glanced at her father and
O'Hagarty, both of whom were talking and
gesticulating at a fearful rate. You showed your
prudence by effecting a retreat, remembering the
old adage, that

> He who fights and runs away,
> May live to fight another day."

"Thanks," said Trelawney, "I accept your compliment, doubtful as it is. Are you so fortunate
as to be acquainted with yonder reverend *charlatan*
—I beg your pardon, Mrs. Ousely—I mean this
Mr. O'Hagarty?"

"We are both of us so far privileged," said
Eleanor, laughing. "We merely turned aside, you
know, to admire the extreme beauty of those
melancholy woods,' and are now on our way to
do the amiable to my father's reverend guest—
despicable renegade!" she added in an under tone,
heard only by Sir James, as they had fallen behind
her mother.

"I quite agree with your flattering encomium!" said Trelawney in the same tone; "what a dull, unmeaning countenance the man has, and yet what vulgar confidence in look and mien. I pity the cause that depends on the advocacy of such men as he!"

Conversátion now became general, and the time passed pleasantly away, till the appearance of John's smiling physiognomy at the door, and his loud, full voice, announcing dinner, put a very agreeable stop to the long-winded account which Mr. O'Hagarty was giving of a great Bible meeting which took place somewhere "away down east."

The dinner went off amazingly well, and, all things considered, very pleasantly. The soup was excellent; the fish exceedingly fine, taken by Ousely himself, as he assured his guests, within twenty-four hours; in short, the dinner was fit to please an epicure, and appeared to give entire satisfaction to Mr. O'Hagarty, whose eyes twinkled with unwonted light as course after course was introduced, and his appetite really appeared to "grow and flourish" as the meal wore on. The feast was, however, anything in the world but "the feast of reason," notwithstanding that there was much loud talk and noisy hilarity, kept up, princi- pally, by the host and his reverend guest, who, to do him justice, was an excellent boon companion. Trelawney was seated between Mrs. Ousely and

her daughter, and whatever " flow o soul" there
was at the table was entirely confined to themselves. When the ladies were retiring, Mrs. Ousely
tapped the baronet on the shoulder : " Mind and
do not stay long here ! join us in the drawing-room
as soon as you can." Trelawney bowed and
smiled assent, and began to meditate a speedy
retreat, looking after the ladies with a sigh as they
vanished through the door. He was not to escape,
however, so easily as he had expected, for when
the wine so plentifully quaffed during dinner began
to work on the brains of the two exemplary
supporters of Episcopal Protestantism, it drew
out some interesting revelations, for which Trelawney was by no means prepared. He had
refused to drink Ousely's toast, consigning the
Pope to warm quarters, whereupon the two
worthies attacked him for his reasons why he
would not drink it. " Because," said Trelawney,
" I consider the Pope as a character to be revered
—as the head of the greatest and most important
association the world has ever seen, and as such
entitled to our respect. And as to his own
individual character, I think Pius the Ninth one of
the greatest and most estimable men of our age.
If I must drink a toast, I raise my glass to him,
the great and good Bishop of Rome."

" Why, what the d—l do you mean, Sir James ?"
cried Ousely, already more than ' half seas over'

—" d) you mean to insult me ?   You are no Pro
testant, sir, if you refuse to drink that toast !"

"No, sir, you're no Protestant !" echoed O'Hagarty.

"Perhaps I am just as good a Protestant as
either of *you*, gentlemen, pardon me for saying
so !" said Trelawney, with his quiet smile. " I may
be a good Protestant, I hope, without dealing out
damnation to those who have never done me wrong.
Now tell me candidly, gentlemen, is either of *you*
a Protestant from conviction—I ask you as gentlemen, as men of honor ?"

"Ho! ho! ho!" laughed O'Hagarty, now
thoroughly fuddled, " I protest that's a good joke.
Now what does the lad mean by a Protestant on
conviction ?   Why, man alive ! there's no such
thing, at least amongst those who go over from
Popery."

"Then what brings them over ?" inquired
Trelawney, carelessly.

"What brings them over, is it ? Why, now,
Mister Ousely, this English friend of yours is
more of a fool than I took him for. Why, my
dear sir—I think they said you were a baronet"—
Sir James bowed—" well, sir, what's that you
asked me—oh! (hiccup) yes, I know—why, sir,
some go over for soup, (and it's none of the best
after all, the blackguards !) some because they had
committed depredations that made them little

thought of amongst the old stock, and some went
for spite—myself for instance !"

" For spite, my dear sir !—how do you mean ?"

"Oh! come! come! none of your questions now
—you see there was a little sly affair found out
on me one fine morning"—he cast a knowing leer
at Ousely ; "and so I found out in my turn that
the Bishop was coming to suspend me, or maybe
worse, so I took leg bail, as the saying is, and came
over to these free and easy Christians who are not
so cursed particular. The Popish religion, sir, is
just like a vice when you're in it—you havn't
room, I mean leave, to turn—you're bound hand
and foot, sir—hand and foot, and soul, and mind--
every little matter is a sin, and a man hasn't the
life of a dog in it. It's an old fashioned religion,
you see, sir, that doesn't make any allowance
for human frailty ; all for the kingdom come,
and nothing at all for this jolly little world of
ours." He then guzzled down another bumper,
and sang in a thick, husky voice :

> " ' They may rail at this life · from the hour I began it,
> I have found it a life full of kindness and bliss ;
> And, until you can show me some happier planet,
> More social and bright, I'll content me with this.'

" Hip ! hip ! hurra !"

" Why, deuce take you, old fellow !" shouted
Ousely, " sure you never told me before the cause
of your leaving Rome, the brazen harlot ! why

your story is something like my own—by Jupiter
it is!"

"How is that, sir? was it spite brought *you*
over, too?" said Trelawney.

"No, no, my lad, I was never brought over. I
was born a good Protestant, for my progenitors,
male and female, were what you might call real
sticklers for the Reformation. Is it I brought
over? I scorn the suspicion of having ever been a
Romanist."

"And you might easily be worse than a Romanist, I can tell you, Mr. Ousely!" muttered
O'Hagarty. "Only for it's being so strict, you'd
never catch me a Protestant. Faugh! a Protestant
indeed—a man might as well be an Atheist, or a
Mahometan, only just that the other is the best
market in this country."

"What's that you're muttering there, O'Hagarty?"

"Oh! some of his old Latin prayers or
incantations," said Trelawney, anxious to preserve
peace. "You were about to favor us with a story
of some kind, were you not?"

"Was I, indeed? what story?"

"You said *your* story was something like that of
Mr. O'Hagarty, if I mistake not."

"Oh! by Jove, yes! I meant that affair of
little Betsy, that went all over the country, I
believe Betsy was the d—l of a fine girl, Sir

James, though she was a sort of a Papist. She
had a confounded old growler of a husband, though,
and when he found out that Betsy and myself
were on good terms, he went straight to the priest.
Well! the priest that was in this parish then--
some five or six years ago, was flaming mad when
he heard of it—he went and spoke to Betsy about
it time after time, but could make nothing of her,
poor faithful creature! so, what did the d——d
old hypocrite do, but he denounced my poor Betsy,
from the altar, and forbid any one to have anything
to say to her, till she'd give up the connection, as
he said. This frightened the poor thing, and she
got so, that if I'd go within a hundred yards of
her she'd run away So you see I lost Betsy, and
was insulted besides, by the interference of that
contemptible Romish priest. But I've had my
revenge, by h—— I had—I swore to put down
Popery as far as I was able, and while there's
breath in my body, I'll do it. Religion has no
business interfering in people's family affairs, and
I'll show Popery that it hasn't, or my name's not
Harrington Ousely."

Trelawney was shocked to hear these revolting
confessions, but he strove to maintain an air of
indifference. ' And how do you succeed, gentlemen, in your laudable efforts to overturn the old
Church ?"

"Not half nor quarter as well as we'd wish,"

cried Ousely, taking the word out of O'Hagarty's mouth.

"And to what cause do you attribute your want of success ?"

"To what cause ? why, to the mulish obstinacy of these Irish Papists—what else ?"

"Ho! ho! ho!" laughed O'Hagarty again with his dissonant voice; "Mulish obstinacy, indeed! by my word, you know little about it. Ho! ho! ho! convert the Irish people, indeed—faith, that's a good notion! Why, Mr. Ousely! you might iust as well think to make the whole of Connemara as level as your table, or—or to wash a blackamoor white !"

Ousely was about to make an angry retort, when Sir James, standing up, proposed to adjourn to the drawing-room, to which the others agreed, after some persuasion.

Let us now return to the O'Daly family, whom we left on their way home, after heaping the last sod on the tender and provident mother, the fond and faithful wife. Granny Mulligan took Eveleen home by the hand, while Phil Maguire and his thrifty wife took charge of Bernard and the eldest girls. There was a fresh outbreak of sorrow when the mourners reached their now desolate home; when they beheld the straw chair in the chimney corner, and thought how she, who for long years had sat in "that old arm chair" was now moulder-

ng in the earth—the warm, loving heart was cold
and pulseless; the mild, soft eyes, that never
looked on husband, child, or friend without a
beaming smile of love, were closed for ever.
These thoughts filled every heart to bursting, and,
for some time, they all sat weeping in silent sorrow, till granny Mulligan, rubbing her eyes with
her blue apron, started to her feet : " Come, come,
children, this 'ill never do—get up now, Kathleen
an' Bridget, an' we'll see about gettin' some dinner.
Tut! tut! Bernard! it 'ud be enough for a child
to cry that-a-way—why, I declare to my sins, little
Eveleen's not one-half so bad. Be off out there,
boys, an' see if Tom Shanaghan's pigs arn't in the
oats! Blessed hour, children! get up out o' that,
and stir yourselves to put the place to rights.
Nanny Maguire, honest woman!" she winked at
Nanny, who well understood her benevolent purpose; "Nanny Maguire, I say! if you go home,
you'll find something to do! it's a shame for you
to be helpin' these children up with their nonsense.
Where's my stick? I'll soon make you all jump!
Eveleen, my pet! did you see that stick of mine?"

With all their sorrow, the young people could
not help laughing to see granny bustling about,
looking for her stick to hunt them, and the kitchen
was quickly cleared.

" Why, blood alive ! granny ! sure you wouldn't
bate us?" said Phil, affecting bodily terror.

"Get out of my way, then," said granny, "an by this 'an' by that I'll lay this cant across your shoulders."

"Oh! murdher! murdher!" cried Phil, "you're a terrible woman, sure enough. Come away, Nanny, honey, or this ould woman will lather us, bad scran to her!"

"That's right," whispered granny, coming up close to the worthy pair; "the sight of you is only makin' them worse, an' they'll do no good while you're here. "I'll be up with you to-morrow or next day, as soon as I see things to rights here."

Bernard roused himself from his sorrowful *reverie*, to "go a piece" with Phil and Nanny, for the young men had already taken granny's advice, and were gone abroad into the fields, to commune together over their heavy loss.

A week passed away, then another, and the grief of the family began to lose its first poignancy. Mr. Ousely had been prevailed upon (through the mediation of his daughter) to grant a few months' reprieve, and, with something like renewed hope, the family-plans were again brought forward, and Cormac and Daniel ventured to remind their father of his promise to let them go to America. The old man was, at first, unwilling to hear the subject mentioned, for his heart was heavy, and well-nigh broken because of his recent loss, but after a few days he began to consider Cormac's arguments, and was forced to admit their justice. Finally, he gave his consent, though, as he did so, the tears were streaming down his furrowed cheeks.

"It's hard, hard," he said, " to part with two of the boys, when the grass is scarcely green yet over their poor mother—the Lord resave her in glory! But sure I know—I know it's all for the best!"

" Be assured it is, my dear father," said Cormac; " if we go now, with the blessing of God we may be able to send you what will help to pay off your arrears, before the first of May—we can do a good deal in nine months, you know, if we get anything worth while to earn!"

" Well! well! I suppose it's God's will," said Bernard with a sigh; " my time here won't be long, children, but if you could redeem the place for yourselves, I'd be well plased, an' very thankful. But what's to be done about the outfittin', Kathleen dear—it falls on you now, *ma colleen!*"

Kathleen looked at Cormac, then at Bridget, and she sighed. She knew of no way of raising the necessaries for the voyage, but she would not grieve her father by saying so. " Well! we'll do our best, father. With God's help, we'll have all ready—but when are you going, Cormac?"

" As soon as I can, Kathleen," said her brother with a melancholy smile, which Kathleen well understood, and so did little Eveleen, too for she said quickly : " If we can only get the things you want—indeed, I hope you'll not get them, so I do —I'll pray every night and morning that you and Daniel may have to stay at home, now mind that, Cormac !"

" Well, but, Eveleen ! my poor child," said her father, " if it's the will of God that they must go, you know we can't have it our own way. If you do pray, dear, say if it's His will to let them stay with us !"

" Oh yes !" said Eveleen, pouting her pretty lips ; " but if I'd pray hard, hard, I'm sure God wouldn't refuse me—doesn't He ever change his mind like us ?"

Cormac laughed, and even Bernard smiled, as he smoothed down the child's silken tresses. " No, Eveleen dear !" said Cormac ; " the decrees of God are immutable—He works out his own wise purposes totally independent of our conflicting plans. Still, we are permitted to ask Him for what we desire, always providing that it be His holy will, or if it be profitable for our salvation."

Still Eveleen could not be convinced but that she ought to pray without any conditions, and she would ask the Blessed Virgin to pray, too, " and then, you know, I'm sure to get my prayer."

" Well ! well ! Eveleen, pray as much as you like," said Bridget ; " only help us to sow some in the meantime—there's the making of three or four

shirts there, that my aunt Biddy sent from Clifden,
so we must all get to work at them, and there's
no one can hem half so well as Eveleen. Come
away, now, dear, and I'll get you something to do."

At this time, granny Mulligan was spending a
few days at Phil Maguire's, according to promise,
and seeing that there was a great hurry of work,
she insisted either on helping Katty Boyce with
the spinning, or Nanny with the knitting.

"Well, then, if you *must* be doin' something,"
said Nanny, "just cast on another pair of stockin's
—we have only the one wheel, you see, an' it's best
to keep Katty to the spinnin', for she's a brave hand
at it."

"Get me the needles, then," said granny, "for I
don't want to be idle. But what hurry are you
in, if it's no harm to ask ?"

"Oeh, *na bochlish*, granny"—she gave her a
nudge with her elbow to say no more. Then lowering her voice, "sure isn't it for Cormac an' Dan
we're hurryin'? Husht! not a word now—Phil
knows nothing about it."

"Don't I, indeed ?" said Phil to himself, for he
overheard the discourse. "*Na bochlish*, Nanny, as
you say yourself!"

"So the three women worked hard and fast for
a whole week, and at the end of that time, on
Friday evening, Nanny tied up half a dozen pairs
of good woollen hose in a bundle, and leaving

18

granny to take care of the house, asked Phil if he wouldn't go down with her to Bernard O'Daly's.

"Why what do you want me for?" said Phil.

"Och, nothing at all, only that it'll be darkish when I'm comin' back, an' you know I'm cowardly in the dark."

"Well! but what's takin' you down now, woman dear?"

"Well, now, arn't you inquisitive?" said Nanny; "don't I want to see the boys afore they go?"

"Are you goin' to bring them anything?" said Phil, with an arch glance at the bundle, plainly visible under her cloak.

"Botheration, Phil! what would I be bringin' them? I can't be always bringin'. Will you come or stay?"

"I'll stay," said Phil, coolly.

"Well! that's all you can do," retorted Nanny, and off she went, her bundle under her arm, all unnoticed, as she supposed. She had not gone far from the door, when Phil was at her side, laughing till you'd think his heart would break, as Nanny afterwards said.

"I say, Nanny, what's that you've under your arm there, like a Hallow-eve goose?" and he laid hold of the bundle.

"Mind your own business, Phil Maguire, an don't be botherin' me. What do you be watchin me for this way?"

"Why, I thought, Nanny," said Phil, still laughing, "that you couldn't afford to give ary more to
the O'Dalys—eh, Nanny! an' that 'the nimble
fingers' had something else to do besides knittin'
for Cormac. Ah! ha! Nanny, I've caught you
this time. Sure I knew well enough, woman dear,
that your bark was worse than your bite. Well,
come along—I b'lieve I *will* step down with you."
They were jogging along very smoothly and quietly
together, when Nanny suddenly discovered that
Phil had a suspicious-looking bulk under the off-
arm, as she said, and she instantly began to bristle
up.

"Arrah then, Phil Maguire, what's that you have
in that bundle?"

"Mind your own business, Nanny," retorted
Phil in her own words, "an' don't be botherin'
me. Step out, woman alive! the night's drawin'
on!"

"I'll not go a step farther," said Nanny, planting her foot firmly on the spot where she stood,
"till you tell me what it is!"

"Hut! tut, Nanny, don't be makin' a fool of
yourself," said Phil, still keeping on his way.

Nanny came up at a brisk trot and placed her
hand on the bundle. "Why then, bad manners to
you, Phil, is it my beautiful web of linen you've
in it? I hope you don't intend givin' that?"

"'Deed an I jist do, then, Nanny!"

" Not a bit of it you'll give them.—I'd see them far enough before I'd give them my beautiful fine web of linen, that cost me a whole winther cardin' an' spinnin'—jist give it to me here now !"

" That's always the way with you, Nanny !" said Phil, keeping fast hold of his prize; " now I'll warrant if I open *your* bundle there, I'll find something else besides the stockin's—you'll give yourself, undher-hand, but you don't want *me* to give anything at all. Didn't I tell you I'd buy you the makin' of a gown if you'd make the frieze —now, I'll keep my word, if you'll only keep a quiet tongue in your head."

" Ay, but it 'id be a good gown that 'id be worth as much as my web of linen. I tell you I'll not give it, now that's all about it."

" Well! well !" said Phil with a heavy sigh, " I see you'll have your own way—there, then, take it home with you, an' I'll go on without it. It's little poor Honora—the Lord be good and merciful to her !—would expect this from Nanny Maguire ! I'm sure, if she could see an' hear what's passin' now, she'd think it was some other body was in it, an' not Nanny Maguire at all ; for she'd say to herself that Nanny Maguire wouldn't grudge the makin' of a few shirts to the poor motherless boys, but it can't be helped !" He was walking on in pretended sorrow, but had not gone far when Nanny was beside him.

"Well! are you comin', Nanny? I thought you had turned back!"

"No! I didn't turn back—don't I want to take these stockin's down, as I have them knit?"

The stockings were given into Kathleen's hands in the course of half an hour, and so was the linen, too, though Nanny slipped it on her knee, and threw her apron up over it. In vain did Kathleen refuse to take such a valuable present; Nanny was resolute, and would not be refused. "An' I'll come down a day or two in the beginnin' of the week," she added, "an' help you to make the shirts. Say nothing about it, though, till I come," and she squeezed her hand impressively. Meanwhile, Phil, though apparently talking with Bernard and his sons, had, by various winks and signs, fixed their attention on Nanny's movements, the kind, though somewhat eccentric creature being too much engrossed to heed them, or to discover that they were watching her with interest.

Nancy Breen, too, brought in her contribution of provisions towards making up the sea-store, and Miss Ousely prevailed upon her mother to send down several articles likely to be useful to the young men. What with one thing and what with another, they were well provided in clothes and all the other necessaries of the voyage: it was only the money that was wanting. This, however was not the least important, and many a consulta tion was held on the subject, b t with little success. It would take six pounds, at the lowest calculation, and that was a very large sum in the present state of affairs. Several of the neighbors were tried, to see if they would lend the money till such times as Cormac and Dar el could repay them, but all in vain. Father O'Driscoll was at length consulted, and at first he shook his head, and his countenance fell. After a moment's reflection, the whole family watching him with anxious eyes, he looked up and smiled: "Well, well, Bernard! we may try to raise the money—the boys must not be detained for so small a sum—and yet it is not so small, either!" he added to himself. "Cormac, could you walk home with me when I'm going? I can get you four pounds at any rate, as Nancy Breen has that much saved with me, and I know she'll be happy to lend it to you."

"Many thanks to you, Father O'Driscoll!" said Cormac, his pale cheek blushing like scarlet, "I'll go with your reverence, and welcome."

Father O'Driscoll and Cormac were scarcely gone, when in came Phil and Nanny Maguire, and with them granny Mulligan, who said she was come to stay till after ' the boys' were gone. Eveleen started to her feet, and running up to granny, threw her arms around her neck, shouting: "We've got it, granny, we've got the most of it; tut 'ndeed, I'd as soon Father O'Driscoll said nothing about it, for now I'm afeard my prayers will be of no use."

"What are you talkin' about, Eveleen?" said granny; "what is it that you've got?"

"Why, the money to pay Cormac and Daniel's passage. We havn't it all, though, only four pounds—father says we want *two* more."

"Husht now, Eveleen!" said her father, fearful lest Phil Maguire might take the hint, and offer the money. "You're too ready with the tongue, daughter dear!"

The visitors put it off with a joke, and then suffered the matter to drop. Phil and Nanny staid for an hour or so, till Cormac got back with the four pounds, and then they hurried away. On the way home, they agreed between them that the poor boys must not be taken short for such a trifle, "especially," said Nanny, "as they'll be sendin' it back by an' bye, an' who knows but they'd be sendin' myself some handsome present into the bargain. Go down in the mornin' with it, Phil."

But Phil's money was not needed, for that night granny Mulligan took Cormac into the room, and taking out an old faded thrash-bag, told him to take what he would find sowed in the one end of it. "There's not much, Cormac dear!" said the kind old woman; "but there's as much as you want now. I was keepin' it to bury me, an' to get masses said for my poor sowl when I'm gone, but

I'll trust to God to give you the manes of sendin it back before I die, an' if yo i're not able to do it, why don't fret about it, *aroon.* Poor granny Mul- ligan has friends enough to bury her dacently even if she hasn't a shillin'. God bless you, Cormac! an' if I die while you're away, I hope you'll pray for me—that's all I want you to do. Not a word, now—I'll be offended at you if you say a word agin takin' what I give you."

Thus interdicted, Cormac could only squeeze the hard, skinny hand held out to him, and, with tears in his eyes, invoke a blessing on the head of his generous old friend—the houseless, homeless wan- derer, with the heart of a princess. Great was the joy of the whole family, Eveleen only excepted, when Cormac announced his good fortune, and it was, indeed, better than he had even anticipated, for the old thrash-bag, when ripped open, was found to contain four gold guineas. Cormac proposed to return the half of it to granny, but she stopped him short, saying snappishly : "Didn't I give you the thrash-bag to keep needles an' thread in, for sowin' on a button or the like? It's yours, I tell you, an' don' be botherin' me any more about it."

On the eve of the day appointed for the young men's departure, Phil Maguire came with three pounds, and was no little surprised to hear that somebody had been beforehand with him.

"Why, where in the world wide did you get
it, Cormac?" he asked in surprise.

"I'll jist tell you that, Phil," said granny, winking at Cormac to keep silent. "There came in a
little ould woman last night, here, an' gave Cormac
an ould thrash-bag not worth a *traneen*, but when
he came to open it, bedad! there was no less than
four goold ginnys in it. Sorra word o' lie I'm
tellin', am I, now, Bernard?"

"Aha!" said Eveleen, "I know who it was!"
and she smiled archly.

"An' so do I, Eveleen!" said Phil. "I know
an ould woman that had *four* goold ginnys these
ten years back, for a certain purpose. Well!
God reward her, anyhow—" he stopped, coughed,
looked at granny's smiling old face; then got up
and shook her hand warmly, and sat down again
without saying another word, but in his own mind
he made a solemn promise, that if God spared
him to outlive granny Mulligan, she should be " as
dacently buried as e'er a woman in the country."

Next day Cormac and Daniel set out for Galway
to take shipping for Philadelphia, being accompanied for several miles of the way by a numerous
escort. Father O'Driscoll had been to the house
in the morning, and gave the young men a letter
of introduction to a priest in Philadelphia, who
had been a fellow-student of his. The two brothers
knelt to get his blessing, and were both cheered

and encouraged when he breathed a fervent prayer
for their success. Bernard and the girls went back
with the rest of ' the convoy,' but Owen and Phil
Maguire, with Larry Colgan and one or two others,
went with them all the way to Galway, nor parted
them till they saw them on ship-board. At part-
'ng, Phil whispered in Cormac's ear, " Don't fret
about them that you're leavin' behind, leave them
to God' an' Phil Maguire, till such times as you
can send them help."

CHAPTER X.

Truth, crush'd to earth, shall rise again
Th' eternal years of God are hers;
But error, wounded, writhes with pain,
And dies among his worshippers.—W C Bryant.

DURING the three weeks that followed the death
of Honora O'Daly, Sir James Trelawney had been
cultivating the acquaintance of Father O'Driscoll,
for whom he began to entertain feelings both of
respect and admiration. Scarcely a day passed
without his seeing the good priest, who, on his
part, regarded the frank and high-minded and generous young Englishman with no ordinary degree
of interest. Father O'Driscoll saw from the first
that his young friend was, like Eleanor Ousely,
desirous of knowing the truth, and solicitous to
understand the Irish people in their relations with
Catholicity. He saw that the prejudices arising
from early and erroneous impressions were gradually disappearing before the increasing light of
truth, aided by assiduous study, but he carefully
avoided any direct allusion to controversial subjects, and never went out of his way to attack
either Protestant doctrines or Protestant practices.
Alone with his God, he prayed earnestly and fer-

vently for the conversion of Trelawney and of
Eleanor; that their minds, already so enlightened
and so well-disposed, might be brought to see the
necessity of joining 'the one fold,' but with them
he never broached the subject, though he met both
very frequently, Trelawney at his own house, and
Eleanor in the cottages of her father's poor tenantry,
while occasionally they all met around the hospitable board of Mr. Dixon.

"The more I see of Father O'Driscoll," said
Trelawney to Eleanor, one evening in the drawing-
room at Clareview; "the more I esteem himself
and respect his religion."

"I told you it would be so," said Eleanor, "for
even I, who have known him for years, can say
the same. His virtues are of that quiet, unpretending kind, which gradually unfold themselves
to our view, and captivate our esteem, nay, our
veneration, without our ever suspecting that there
is anything remarkable about the man."

"For my part," said Sir James, "I consider
such a man as the greatest blessing in society;
heart and soul devoted to the good of his fellow
men, with the grand ulterior view of promoting
religion and the glory of God; pursuing 'the
calm, unbroken tenor of his way' through good report and evil report, without any of those earthly
ties which bind the heart to this world; devoting
the greater part of his small income to the relief

of his suffering flock, as I know he does; oh! surely, Miss Ousely! such a man as this cannot be the minister of a corrupt and corrupting Church!"

"And who said he was, Sir James?" said Elea nor, smiling at his generous warmth; "I am sure I never did. Why, my dear sir, if the Catholic Church be not the religion of Christ, it has, then, disappeared from the earth."

"Am I to understand that you mean the *Roman* Catholic Church?"

"Certainly, Sir James, I mean no other. There was a time when I fondly imagined that our Church of England *was* a branch of the great Catholic Church, but I have since studied the matter by the light—not of reason alone, but of reason cou pled with Scripture and Ecclesiastical History, and I have come to the conclusion—I trust through the mercy of God, that the Roman Church is the only ark of safety amid the deluge of corruption which covers the earth."

"And did you arrive at this conclusion without any outward agency?" asked Trelawney, more and more struck by the extraordinary power of Eleanor's mind.

"Not exactly," said Eleanor; "my aunt Ormsby —of whom I have sometimes spoken .o you—has lately become a convert to Catholicity, and her letters have expedited my progress no little.

19

Neither my father nor mother yet knows of her conversion, but I know it, and God knows it," she added with touching fervor.

"It is a remarkable fact," said Trelawney, musingly, "that the converts to the Catholic Church are generally, I might say nearly always, from amongst the educated classes, while those who go forth from her communion are the unlearned—the poor—"

"The starving, Sir James! allow me to suggest a word. The reason of this difference is very plain. The Catholic Church employs no direct means to gain converts. She prays for the conversion of sinners, infidels and heretics taken collectively; she edifies the world by her admirable and ever-renewed works of charity; she silently presents to us the perfection of Christian life, exemplified in her monastic orders, and in a vast number of her secular clergy. The rest she leaves to God, knowing that He only can touch the heart, and draw water from the hard rock. Hence it is that her converts are those who have time and opportunity to read and to *think*. As for the converts from Catholicity, why—the less we say the better it is. They are, for the most part, poor starving creatures, brought over, like the apothecary in Hamlet, because of their necessities, to sacrifice the hope of heaven for the more immediate prospect of preserving their wretched life here or earth They are, in nine cases out of ten, the most miserable, the most ignorant, and the most worthless of the community, and the exceptions are scarcely more worthy of respect—the apostacy of a priest is the greatest triumph ever obtained by the proselytizers, and of that unhappy class, you have a very fair specimen in my father's bosom crony, Mr. O'Hagarty. I have seen several individuals of the species, and I can solemnly assure you that such is the case; meet an apostate priest where you will, and you will find him stamped with sensuality, gross selfishness, rabid vindictiveness, directed against the Church which he had disgraced by his ministry."

"In the same way," said Trelawney, "that Satan and his rebel angels are the most inveterate haters of God, and would fain debar all mankind from that heaven which themselves have lost for ever— a very natural feeling, all things considered."

"What's going on now?" said Amelia Dixon, a light-hearted, happy-looking girl of eighteen or nineteen, as she threw herself on the sofa beside Eleanor; "I really think you two are plotting some mischief—take care that we do not find some vile Meal-tub Plot coming to light one of these days; you the Titus Oates, cousin, and my sedate friend, Eleanor, the—who—oh! ye stars, help me to a name!--what a pity Nell Gwynn

wasn't a party concerned—your name would just
suit, you know, Nell!"

"You are exceedingly kind," said Eleanor,
laughing; "but as you are rather unfortunate in
your historical allusions, don't trouble yourself
ransacking amongst the debauched men and women
of the Merry Monarch's court, for comparisons
which *might* be invidious. Just tell Sir James
how you came to give up those notions which you
had a year or two ago, about converting the Papists. Perhaps, though, you have told him already?"

"No, indeed, not a word of it—he laughs at me
now for all manner of odd blunders, as he says,
and I should be very sorry indeed to give him such a
grand subject as that—I don't like to have people
laughing at me!" said Amelia, pouting, and looking as though she were half inclined to cry, though
there was a 'laughing devil' lurking in the corner
of her bright eye, and certain dimples playing
around her small mouth, which showed her more
disposed to laugh than to cry.

"Come, now, my pretty cousin," said Sir James,
"forget and forgive—I promise beforehand never
to turn your confession to account against you. I
should like, of all things, to hear how *you* came to
think of proselytizing."

"Why, that was not the strangest part of it,"
said Amelia briskly; "we had a governess just
then, who was brimful of the notion—my stars

how she would extemporize on the horrors of Popery, and on its baneful effects, social, political,
and religious! At first, we children used to laugh
at her over-drawn pictures—caricatures they were,
in reality, of what she was pleased to call *the
great superstition*—sometimes *the great delusion*—
by way of a change, you know ; but gradually we
began to listen with more interest, whereupon
Miss White poured forth her harangues with still
increasing *unction*—(isn't that the word, Eleanor?)
and what with that and the peculiar nature of our
studies—as ultra-Protestant and anti-Popish as
could be—we got our heads full of the romantic
notion of besieging the citadel of Popery in right
down earnest. Bless you, cousin James! we were
filled with what Miss White called ' holy zeal,' and
our pocket-money, for many a long day, went into
the coffers of Bible Societies, and the Church Mis-
sionary Society, and the Tract Society, and—oh!
dear! how anxious we were to do something for
the good cause. Mamma, and papa, and Arthur,
and all the rest, used to laugh at us, but we didn't
mind—we only pitied their blindness, and kept
hoping for their speedy conversion. At last we took
it into our heads to try our hands amongst papa's
tenantry, and so, having provided ourselves with
ever so many tracts of the most unctious and persuasive kind, we went to work with the most
sanguine hopes of success. You should have seen

Eliza and myself, marshalled by Miss White, tramping about from cottage to cottage, with our reticules and our pockets stuffed full of tracts, with perhaps a Testament or two in the hands of each of us, just by way of sign-post, to denote our godly avocation. La me! what a figure we must have cut!"

Trelawney laughed heartily, and Eleanor smiled. Amelia affected a gravity all unusual to her, and sat waiting very demurely for her cousin's mirth to subside. When his features were again somewhat composed, she turned to him very coolly:

"Well! Sir James, are you done? I thought you were not to laugh at me any more—eh?"

"Why! you little mischievous elf, how could I help laughing—but you have the best of it this time—and that you well know—you do all you can to excite our risible faculties, and then blame us for falling into the trap—'Ah! little traitress!'—you know the rest. Pray, proceed with your narrative—how did you fare amongst the peasantry?"

"Why, not very well, I must confess!" said Amelia, with well-feigned reluctance. "The result was not quite what we expected."

"Well, but why not tell all, *cara mia?*" said Eleanor.

"What do you call *all*, Eleanor?" Then turning to Trelawney –"The whole amount of it is.

cousin, that we thought we had nothing to do but make our appearance, Bible or tract in hand, and that, *presto*, the whole phalanx of Popery would whisk off out of our sight in a flash of fire—*veni, vidi, vici*, was the real motto of our warfare, though to be sure, we didn't say so even to each other, but lo! and behold! the stout old body corporate not only resisted, but actually got the better of us, and that without an effort. It is a great old institution—that same Romish religion!—this in parenthesis, cousin! Well, *seventeenthly*, as old Mr. Fumbleton says, about the middle of his sermon, we had only made a few proselytizing visits when I was glad enough to slip my Testament into my pocket, and to tell you the truth, I was profane enough to throw a dozen or so of tracts into the brook about the same time. Eliza held out a little longer than I did, but even she gave up in despair after a few weeks, whereupon our saintly duenna was fain to desist—' for,' said she, in the grief of her heart, ' since you will not accompany me any more in my visits of charity—why, I fear I must discontinue them altogether, for there is no saying what these miserable Romanists might do if they caught a young lady alone—even as it is, I find them anything but civil.' So ended our campaign against Romanism, to the infinite amusement of papa and mamma, who used to joke the three of us so unmercifully that poor White's *evangelical* temper could not bear it, and one fine morning she tendered her resignation, which was thankfully accepted."

"Alas, poor Yorick!" sighed Trelawney, with affected sympathy. "But how did the people receive you on those occasions? I should rather ask how it was that you became so speedily disgusted with your self-imposed task."

"Why, bless your dear simple heart, cousin, can you not see the reason? Have you ever talked with the peasantry on religious matters?"

"Not exactly on *religious* matters?"

"If you had, you would need no explanation, that's plain; why, cousin, I saw from the very first (though I left the conversation principally to Miss White), that the poor simple cottagers, ignorant as they were in other matters—had more correct notions of religion, in all its essential doctrines, than we ourselves had. I actually felt ashamed when I heard them give such clear, intelligible answers to questions which might have puzzled more learned men and women. It was amusing, however, to hear Miss White cross-examining some of them. 'Poor people!' she would say, 'I pity your condition—your minds are so darkened by the gloomy shadow of Popery. Now if you would only believe in the Lord Jesus'—'Stop there, ma'am—stop there—sure we do b'lieve in him— oh, bedad, we do so—the Lord pity us if we

didn't!'—'Oh! but then, you put too much trust
in the Virgin; you know ¬r ought to know that the
Romish Church makes a goddess of her, and prays
to her as such.'—'Beggin' your pardon, that's not
thrue. I hope you'll excuse me, ladies—sure,
ma'am, we only ask her to pray for us, an' that's
not the way we spake to God, you see. Oh! be-
garra, ma'am, it's not the same case at all—sure
every child knows that.'—'Well! well!' Miss
White would say, a little disconcerted, ' but then
look what nonsense it is to pray to the saints—
what can they do for you?'—'Well, ma'am, if you
don't like to ask them to pray for you, why, the
loss is your own—but it's our notion, ma'am, that
it's a fine thing all out for poor sinful creatures like
us to have sich friends in heaven to put in a good
word for us—an', sure enough, we think it's no
throuble to ask them to do it too. You might do
worse, lady an' all as you are, than ask them to
pray for yourself.' On hearing this, or some such
answer, Eliza and I would burst out laughing, and
Miss White would flounce out of the house, muttering all sorts of execrations against what she
called the obstinate folly of the Romanists. But
here's Arthur with his violin—let us have a dance,
Eleanor! Your mother and my mother are lost
in the mysteries of backgammon in the corner.
There cousin, take Eleanor's hand—no excuses
now—I'm going off to hunt up Eliza and Ed-

ward." Away she ran, leaving Eleanor and Tre
lawney once more *tête-a-tête.*

"Take Eleanor's hand!" repeated Trelawney,
fixing a keen glance on Eleanor, whose eyes sought
the ground; "that were too much bliss for me,
what a world of happiness is comprised in tha
little word, so lightly spoken!" There was an
earnestness in his tone, which arrested the smile
that was hovering on Eleanor's lip, and brought
the warm blood to her cheek. But her presence
of mind never forsook her. In a moment she was
calm and composed as usual, and said, without ap
pearing to notice the words just spoken :

"We were talking of Father O'Driscoll, Sir
James, were we not? This episode was rather a
long one !—you have, doubtless, made the acquaintance of some of his flock—what do you think of
the O'Dalys?"

"They are a most estimable family!" said
Trelawney, making an effort to imitate Eleanor in
her graceful self-possession. " I have talked a good
deal with that young man, Cormac, who has just
left for America, and I found him possessed of
much solid information on very many subjects,
the whole based on thoroughly religious principles."

"You will always find that characteristic in
Catholics who have been subjected to purely
Catholic training. With them religion is at the
bottom of everything—God the *Alpha* and *Omega*

Religious instruction is, consequently, the primary object in Catholic education, while secular learning holds but a secondary place, coming in only as an accessory. From the cradle to the grave, the Catholic—that is, the true Catholic—makes religion the one grand affair of life, and yet he fulfils all human duties with a cheerfulness and a readiness that contrasts well with the cold, proud spirit of Protestantism. There is no parade or ostentation about Catholic charity, as you must already have observed."

"Witness Phil Maguire," said Sir James with a smile.

"Yes, and his wife Nanny, who, with all her apparent *closeness*, is, at bottom, no whit behind Phil in generosity, or rather charity. I have a great respect for both of them, and it does my heart good to see them trudging along together, like Darby and Joan, on Sunday morning, dressed up in their best, the very picture of contentment and good nature. Ah! I am sure—sure that God looks graciously down on that worthy couple, with all their quaint eccentricity of manner, for they are covered with alms-deeds and good works. I would rather be one of those kind-hearted, simple peasants, praying before the altar of sacrifice in their lowly chapel, thanking God alike for the good things and the evil things which he sends them, than the highest and mightiest of our Protestant

magnates, odious before God and his saints because
of their foolish pride, and hypocritical pretences,
and stony hearts."

"Why, Miss Ousely, you speak warmly on the
subject!" observed Trelawney, his own cheek glowing and his eye flashing with something of Eleanor's
excitement. "You speak of the Irish peasantry
in a very different way from that in which they are
represented in Exeter Hall!"

"I do," returned Eleanor, still more warmly
than before, "because I *know* those of whom I
speak, and have no interest in caluminating them.
I have seen them in all the various circumstances
of life—I have stood by their death-beds, as we
both did at Honora O'Daly's, and I tell you, Sir
James Trelawney, that I have long ago learned to
reverence their virtues—and the religion by which
those virtues are fanned into warmth. Very often
have I felt myself ready to bow down before some
poor, half-starved man or woman, sitting lonely and desolate in the cold, bare cabin, when
amid all the privations of their lot, they would
raise their eyes to heaven, and say : 'God's will
be done!' and then, when I went forth from that
scene of misery and of heavenly resignation, it
has often been my lot to meet the Scripture-reader,
McGilligan or such as he, going in with his bundle
of books, to mock the sufferings of the unhappy
inmates with the offer of a tract or a Testament;

and if you told them of the utter destitution of
the poor creatures, they would turn up the whites
of their eyes and groan out : ' Alas ! if they would
only read this blessed book, and believe its glorious promises !' Ah, Exeter Hall! Exeter Hall!
have I said to myself, these are thy agents—not a
mouthful of bread for the starving, but plenty of
tracts and Bibles. But how I am forgetting
myself!" she suddenly added, seeing the earnestness with which Trelawney listened. "You, who
are a stranger, cannot understand these things, or
enter into my feelings."

"I can and do understand, Miss Ousely—I have
studied both sides of the question theoretically
within the last few months, and practically within
the last few weeks, and therefore"—

"And therefore, you know the difference, I suppose,

' 'Twixt tweedle-dum and tweedle-dee.'

"There, now, fair lady and fine gentleman, as I
have decided the matter from that reverend authority, Hudibras, you must e'en give in, and come
along. We're going to get up a set of quadrilles
outside here."

There was no getting over Amelia's off-hand
brusquerie of manner, especially as she laid hold
of Eleanor's arm on one side and Trelawney's on
the other. So they laughingly gave themselves

up, and marched off right willingly with their fair captor. By the time the young people had got through their set of quadrilles—the "Lancers" I believe they danced on that particular evening—the elder ladies had finished their game, and Mrs. Ousely ordered her carriage.

"I thought Mr. Ousely was to come for you?" remarked Mrs. Dixon.

"His promise was only conditional," said Mrs. Ousely, "and I suspect he has his new friend, Mr. O'Hagarty, who often drops in of an evening to discuss religious matters, and—".

"And drink whiskey-punch, mother," said Eleanor. "He would have made an admirable priest of Bacchus, had he lived in Pagan times!"

"The horrid old bore!" exclaimed Amelia. "I can't endure him—he stares one out of countenance. I think the Church of Rome showed its good sense, aye, and good taste, too, in getting rid of that fellow. I'm very sure that he is no great credit to any religious body, for the man looks as though he were half stupefied with drink. Faugh! such converts as we have! I'm sure they're not worth one half what they cost, and, for my part, I think it's very mean of the Church missionaries to have any thing to say to them—belly-friendship is poor friendship, and they'll be all going back again to the old Church when times are getting better!"

Mrs. Dixon laughed at her daughter's lively
sally, for she herself had no sort of sympathy with
the Jumpers, but seeing a cloud on Mrs. Ously's
brow, she gently changed the subject, and made a
sign to Amelia to desist. Sir James and young
Dixon proposed to accompany the ladies on horseback, and their offer was, after some hesitation,
accepted.

Next day, about noon, Trelawney rode up to
Father O'Driscoll's door, and asked the housekeeper whether the priest was at home.

"Well, no, sir," said Nancy Breen, making a
low curtsy; "he's just gone down to the poor-
house about some little orphan girls that they're
for makin' Prodestans of among them. Bad scran
to them for schamin' villains, if we haven't the time
of it with them, one way an' the other. Beggin'
your pardon, sir, for sayin' the like before you,
that's one o' themselves."

"I deny it, Nancy," said Trelawney, laughing, as
he threw himself off his horse. "I never did or
never will belong to a fraternity to which good
people must apply such epithets—I have nothing
to do with your 'scheming villains,' as you call
them—and not unjustly. If you will permit me
to sit down till Father O'Driscoll comes, I shall
take it as a favor. Here, boy, put my horse in
the stable."

Nancy ushered the visitor into the priest's little

sitting-room, and having stirred up the turf fire in
the small grate, she closed the door, and quietly
withdrew. Sir James had just taken down a
volume of St. Alphonsus Liguori's *History of the
Heresies*, and was just turning over in quest of the
great English revolt—commonly called the Refor
mation, when his attention was arrested by a man's
voice, talking with Nancy in the kitchen outside.

"Come by an' sit down, Shane," said Nancy;
"what's the best news with you?—good news is
scarce with some of us these times!"

"Why, then, indeed, Nancy *ahagur*, I've the
best news I could wish to have, thanks be to Him
above!"

"Ah, then, do you tell me so, Shane? an' what
is it, *agra*? Is there any account from beyant the
wather?"

"'Deed an' there is, Nancy, 'deed an' there is
then. Look there! that's what I got in the post-
office yesterday—and look there again—see what
was in it!" He hastily opened the soiled and
badly-directed letter, well-nigh covered with post-
marks, and took out a bank-post bill for *ten*
pounds, holding it up before Nancy's widely dis-
tended eyes.

"Why, dear bless me, Shane Finegan! sure you
didn't get all *that* from America—if you did, you're
the lucky man all out!"

"Faix, an' I got it jist as you see it, Nancy!"

returned Shane, somewhat proudly. "I always
knew an' said—even when nobody b'lieved me—
that Johnny an' Susy wouldn't forget their ould
fat.er. But is Father O'Driscoll at home—I want
to spake to him." There was a suppressed exulta-
tion in the poor old man's manner and in his shrill,
feeble voice, that was in strange contrast with his
ragged habiliments and poverty-pinched features.

"He's not at home, then," said Nancy; "he's
down at the poor-house. Is there anything wrong
at home?" Nancy's curiosity was thoroughly
aroused.

"Nothing wrong, thanks be to God, but everything right: I'll jist tell you a secret, Nancy, for
you're an ould friend; sure it's Harry that sent
me over to spake to the priest."

"Arrah no, then, Shane! is it thruth you're
tellin'? Sure we counted Harry as good as lost,
bekase he went roun' with the thracts, you see."

"Ay, but that was only to make sure o the
bread for me an' the ould mother at home. To be
sure it was a great risk for the boy to run, bekase
he might be out off any minute, an' him in *that*
state; but, you see, Nancy *agra*, there was no
earnin' to be got, an' poor Harry couldn't bear to
see myself an' his mother sufferin' hunger an'
starvation, so he gave in to them for a while, until
sich times as we'd get something from America.
But he's overjoyed now, poor fellow, that he can

snap his fingers at the Jumpers, an' pitch their
soup to the divil where it came from, Lord save
us ! The first thing he said, when he read the letther and seen the money, was : ' Now, thank God,
I can get into the ould ark again !' an' the first
thing he did was to take the bundle of thracts he
had in the house, an' fling them into the fire. Oh !
maybe myself an' Molly didn't laugh au' cry with
joy, an' it was for which of us would hug Harry
the first, when we seen the unlucky papers blazin'
—an' if Molly didn't stir up the fire about them
it's a wonder—faix, Nancy dear, she stood watchin'
them with the tongs in her hand, till they were all
in a cindher, an' then she put down the tongs with
a clash, an' said, ' The Lord in heaven be praised.' "

"Why, then, Shane Finegan !" said Nancy,
wiping away the sympathetic tears which bedewed
her cheeks ; " why, then, myself's well plased to
hear this news—it used to go to my heart to hear
Harry Finegan called a Jumper, an' to think of
the shame an' the sorrow that he brought on your
self an' Molly—God knows I offered up many's
the Pather an' Ave for his comin' back agin. Will
you take a dhrink o' milk, Shane ? it's the only
thing I have to offer you. She lowered her voice,
" You know it's not ould times with us, Shane—
there's neither money nor anything else comin' in
now, barrin' what Phil Maguire an' a few others
sends us, an' even that we can't call our own, for it

it was the last bit or sup that was in the house,
it'.l go when any one comes makin' a poor mouth
to his reverence—an' och, och, Shane! but that's
often enough, God help the cratures that has to do
It. I don't know what in the world I'd do to keep
the house a-goin', if it wasn't for the cow an' the
few hens that I manage to keep."

"'Deed, then," said Shane, "you make your
milk an' butther, an' eggs go a long way, for I
never knew any one to come askin' the bit an' sup
from you, that hadn't it to get. The Lord be
praised!"

"You may well say that, Shane!" returned
Nancy; "the little that's in it goes a great way
among the poor—an' sure that's no wondher, when
we think of the five loaves and three little fishes
that Our Saviour multiplied till they fed five
thousand people. Nobody'll ever go hungry from
Father O'Driscoll's door—mind I tell you *that*,
Shane! for charity's in his heart, you see, an' he'll
never be left without the manes of showin' it."

Nancy's allusion to the loaves and fishes struck
Sir James as something strange, but in the course
of his after intercourse with the peasantry, he be
came aware of the fact that they are far from
being ignorant of the Scripture.

While Sir James was still reflecting on Nancy's
words, he heard Father O'Driscoll's voice outside,
accosting Shane.

"Why, Shane Finegan, is this you! I have not seen you for a long time. I suppose Molly told you that I called two or three times?"

"Och, musha, then, she did so, your reverence," said Shane, standing up, "an' one o' the times I was in the little room within, but was afeard an' ashamed to come out, on account o' poor Harry's bad doin's. But now I can hould up my head, an' show my face agin, thanks an' praises be to God— the heavy load is taken off my heart, an' I'm a new man altogether."

"Why, Shane, I really believe you *are* a new man, as you say yourself. What is the cause of this sudden change? Has Harry come to himself again? Nancy," he said, in a low tone, "did you give Shane something to eat?"

"Well, no, your reverence, I did not—but I gave him something to drink."

"Pooh, pooh, Nancy, go and get him a bit to eat—he has travelled a good way this morning." Nancy disappeared.

"You were asking me about Harry, your reverence," observed Shane, "an' what makes me so joyful this mornin'. Sure we got a letter from Johnny and Susy yesterday, an' *ten pound* in it— sorra penny less, your reverence, an' as soon as ever Harry read the letther, by the laws he jumped two feet from the flure, and made a dash at the thracts, where he had them, up on a shelf, an pitched them into the fire, an' was as joyful as ever you seen any one in all your life."

"Thanks be to God, Shane! thanks be to God!" said Father O'Driscoll, with pious fervor. "I never lost my hopes of Harry, for whenever I chanced to meet the poor fellow he always tried to avoid me, and I could see that conscience was busily at work within him. It was only stern necessity that induced him to do what he did. I always pitied, more than I blamed him—his crime was grievous, but not altogether inexcusable."

"It was as good as a play, your reverence," said Shane, "to see Andrew McGilligan, when he came this mornin' to get Harry to go up to Dan Leary's with him."

"Sit down, Shane," said the priest, as he took a seat opposite, "and tell me all about it. It must have been amusing, for I know Harry has a great deal of humour."

"Well, your reverence, we got a stone of meal from Barney Flynn till we'd get the draft changed. We didn't like to brake it till your reverence 'id see it, an' we were jist afther our breakfast, when there came a long shadow over the flure, and when we looked up, bedad there was Andy at the door, as large as life, an' as sour as vinegar. He never put the spake on Molly or me any time he came, for he had thried it in the beginnin', and got some ill-sarved answers that didn't plase him, but he says to Harry, 'Are you ready to come with me ?'—'No!' says Harry, as short as could be—'I'm not goin'.'.'— Not goin',' says Andy, 'and why not?' says he.—'Because I'm not goin' to act the hypocrite any more,' says Harry. With that you'd think that McGilligan's big eyes grew twice as big, an' he looked at Harry as if he'd look him through.—'Why, what do you mane, Finegan?' says he.—'I mane jist what I said,' says Harry, 'I took my turn out o' you, an' got what I wanted, an' I thank you kindly for helpin' me along through the bad times—though to be sure you didn't do it for charity, only bekase you thought you had me bought body and soul.'—'Why,' says Andy, 'sure it can't be that you're goin' back to Popery?'—'I'm not goin' back,' says Harry, 'for I never left it—God forgive me for makin' fools of so many wise men—but it wasn't my fault—you might blame yourselves. You're always tryin' to buy up consciences, an' you oughtn't to complain when you find people playin' sich ugly thricks on you.'—'Where's the tracts?' says McGilligan, scarce able to spake with anger.—'In the fire there,' says Harry back again, 'where they ought to be. We made a bone-fire of them.'—'Very well,' says McGilligan, 'that'll do. I suppose you've got some money some way or other,' says he, 'when you're gettin' so saucy, an' you may depend we'll put you through some of it before we quit you

We'll make you pay for the tracts, and a trifle
more, too.   You'll not get off so easy as you think.'
With that Harry lifted this stick o' mine that hap
pened to be in the corner beside him, an' he made
a flourish as if he was goin' to sthrike the Bible-
reader, though he was only playin' a trick on him,
bekase he knew what a coward he had to dale
with.   'By this an' by that,' says Harry, 'if you
don't make yourself scarce, I'll give you the
weight o' this.'   He hadn't to spake twice, for
before you could snap your finger, there wasn't a
color of Andy to be seen, an' you'd think Harry
'id jist brake his heart laughin'."

"It was certainly a summary way of getting
rid of him," said Father O'Driscoll with a smile.
"But I fear Harry will find these people very
troublesome—they are not apt to forgive, and can
do much harm."

"Oh! as to that, your reverence, Harry doesn't
care a fig for them—he's comin' up this evenin' to
see you, but he was ashamed to come near you,
till you'd hear how the matther stood.   Don't be
too hard on him, your reverence, for it was love
for me an' his mother that made him do what he
done, always hopin' that God 'id give him time to
repent an' do what was right."

Father O'Driscoll shook his head, but thought·
it unnecessary to continue the subject, so he mere

ly assured Shane that he would receive his son kindly, and then passed on into the room, Nancy having informed him that ' the English gentleman was there.

After the usual friendly greeting, Sir James referred to the conversation which he had just heard, and gave Father O'Driscoll an account of the first part, with the exception of that which related to his own affairs. "There were many points of interest in that conversation, simple as it was," said he ; "points to be remembered and dwelt upon in days to come. Leaving apart the main subject of the young man's return to Popery, as McGilligan said, what chiefly struck me was your good housekeeper's introduction of a certain passage of Scripture. In England, it is commonly believed that Papists, especially the lower orders, know nothing whatever of Scripture."

"Well, my dear sir !" said the priest, "I can only say that those who think so know nothing of us or our people. You will find, if you take the trouble of examining for yourself, that even the most illiterate Catholics have a certain knowledge of the principal events recorded in Scripture, especially in the New Testament, and make a better application of what they know than very many of your Bible-reading people. It is because, instead of giving them the Bible to con over, we explain it for them, and teach them to regulate their lives by its precepts. But I must leave you for a moment till I see what Nancy has got for dinner—if she has anything eatable. you must stay and dine with me "

CHAPTER XI.

What war so cruel, or what siege so sore,
  As that which strong temptation doth apply
Against the fort of reason evermore,
To bring the soul into captivity ?
                              Spenser's Fairy Queen.

To trample on all human feelings, all
Ties which bind man to man, to emulate
The fiends, who will one day requite them in
Variety of torturing.
                              Byron's *Two Foscari.*

A FEW more weeks passed away, and through
the kindness of Phil Maguire and his wife, together
with what Owen could earn—it was very little, for
there was scarcely any work to be had—poor
Bernard O'Daly and his children were enabled to
live. The dull, damp days of October were
nearly past, when the Ousely carriage rolled ra
pidly along the principal street of Killany one
morning about nine o'clock. Mr. Ousely himself
was in the carriage with his wife and daughter,
being on his way to the court-house, for it was
law-day, and he was, of course, a J. P. The ladies
had some shopping to do in town, and also a few
visits to pay, amongst others to the lady of the
officer in command of the detachment then sta-

tioned in Killany Barracks. As the carriage passed through the market-place, there was a crowd of men standing there—gaunt, hungry look ing creatures, with tattered, or at best thread-bare garments, their limbs shivering with the cold, and their benumbed fingers scarcely able to hold the spades whereon they rested; they were laborers, waiting for employment, of which there was little chance at that advanced hour of the day. As Eleanor glanced over the motley crowd, consisting of all ages, from the old man, well nigh past his labor, to the stripling, who was scarcely fit to un-dertake a day's work, her heart sank within her as her eye fell on the handsome face—no longer ruddy—of Owen O'Daly, where he stood a little apart from the rest, leaning against the corner of a house—one hand thrust into the bosom of his thin linen jacket, and the other grasping, as it were convulsively, the handle of his spade. An old cloth cap was drawn down close over his brow, and his fine hair hung dishevelled around his tem ples, while his eye-brows were knit almost together, and his eyes had a strange, scowling look, that made Eleanor start. Alas! how unlike the laughing, light-hearted boy that he had been but one short year before.

"Oh, mother! mother!" said Eleanor, in an un lertone, "do look there!—Is it not pitiful to see that poor lad, Owen O'Daly, standing there,

in such a condition!—my heart aches for him!—
It is bad enough to see any of those poor men,
and to think that they have been waiting there
since early, early morning—but, oh, mother dear!
it is grievous—grievous to see that young O'Daly
there—he whose prospects were so bright but two
years ago!"

To do Mrs. Ousely justice, she was quite as
much shocked as her daughter, but her husband
had no pity to throw away on such a subject.
"See what a scowling look the fellow has!" said
he. "He bids fair to become a regular despera-
do!—I should not wonder to hear of him taking
aim at some one from behind a hedge!"

"And little wonder if he did!" thought Eleanor,
but she wisely refrained from saying so, fearing to
irritate her father.

"It's altogether their own fault," said Ousely,
working himself up into a sort of passion. "If
they would only do as they ought to do, that d——d
young scape-grace needn't be standing there like
patience on a monument! They're getting another
chance to-day, and by ——, if they don't give in,
out they go, if they were O'Daly a thousand times
over!"

Eleanor looked inquiringly at her father, but he
seemed determined to give no explanation, but
kept nodding his head, and muttering to himself,
and looking out of the window with a frowning

aspect, as though to deter those within the carriage from any attempt to penetrate his meaning.

Meanwhile, let us return and see what was going on in the now desolate homestead of the O'Daly's. It might be ten o'clock in the forenoon, when Andrew McGilligan, and another Scripture reader, named Timothy O'Hanlon—(or *Hanlon*, as he latterly styled himself, in holy horror of the old Milesian O')—made their appearance, demanding if Bernard O'Daly were at home. Kathleen replied in the affirmative, and sent Eveleen down to the room for her father. The old man started, and his pale cheek was flushed for a moment, when he saw who his visitors were. Still mindful of the old hospitality of better times, he first invited the men to be seated, and then asked what they wanted with him.

"Old man!" began McGilligan, " we have come again to seek the lost sheep of the house of Israel, and to offer you once again the word of salvation, the true bread of life! yea, we are grieved and sorrowful to see the misfortunes which have befallen you, and would rejoice to apply a remedy if you would only let us!"

"My misfortunes are from God." returned Bernard, slowly, "and the remedy is not in your power to give. I'm willin' to bear whatever thrials God sees fit to send me."

"It isn't the will of God that makes you poor

and miserable as you are," said Hanlon, suddenly
breaking silence; "it's your obstinate attachment
to Popery—that's what does the mischief, and
your priests put that cant about God's will in
your mouth, so that you may deceive yourself and
others.   Come, now, be wise for once in your life,
and listen to us!"

Bernard O'Daly stood up, the fire of his young
days, the fire of his Catholic faith, flashing from
his eye; his cheek glowed with a hectic flush, and
a strength which he had not felt for years gave
energy to his words and manner : " Get up an' go
your ways!" said he, pointing to the door; " you
say I'm poor an' miserable, an' so I am, God
knows, but this house is still mine, an' old as I am,
I'm able for two such *leprahawns* as you any day,
so go at once, or I'll send you out head-foremost!"

" Father dear!" said Kathleen, coming forward
from where she and her sister Bridget were *quilt-
ing* at a frame in the farther end of the kitchen;
" Father dear! don't mind them, let them go
quietly!"

" Oh, surely, miss, surely!" said McGilligan in
an ironical tone, " he'd best let us go quietly, as
you say.   But it's only proper and christian like
to let you and him both know that our visit is the
last chance he'll ever have, if he now holds out
against the religion of Christ."

" Don't dare to blaspheme in my presence!"

ried Bernard, sternly; "that's worse than all. You came to insult myself an' my children with your sham of a religion, when there was death an' black sorrow in the house I havn't my brave boys now to give you the door, but even so, you'll not brow-beat me with your threatenin'. Get you gone, I tell you once for all. Let me alone, children, I'll do them no harm—I don't want to do them any, bad as they are—all I want is for them to leave my house!"

"*Your* house!" cried Hanlon, scornfully; "it's as much yours as it's mine, and maybe far less! Come away, Andrew! let us leave the old reprobate to his fate—even as Israel, the adulteress, would not hearken to the prophets, nor give up her fornication, till the wrath of God came down on her in a boiling stream, so shall this hardened sinner be burned up with all that is his! Ah! we shall see it, our eyes shall see his utter ruin, and that before many hours are passed!"

But McGilligan would make another attempt: "Bernard O'Daly!" said he, getting near the door, however, as he spoke; "Bernard O'Daly! think of your children—behold those comely young maidens; they are poorly and meanly clad; they and you have known hunger and want—will you see them starve, as many others have starved! oh, be merciful to your own flesh and blood' If you now reject our offers, we are authorized to say that you and yours shall be turned out on the road to starve and die!"

"Better that than endanger our souls!" said Bernard, resolutely; "we can get over all that so long as we have the thrue faith, an' if we hadn't it, all the riches in the world wouldn't be worth a pinch o' snuff. My children an' myself are in the hands of God, an' we disregard all you can do! That's the last word, now—go back with it to your employers. Tell them that the O'Dalys are of the ould stock, or the ould rock, your choice, an' they can die *for* their faith, as they have lived *in* it, them an' their fathers before them."

"Very well, then," said Hanlon, "you needn't blame us for what's to come."

"We go," added McGilligan; "but we go, shaking the dust from off our feet, like the Apostles of old." Bernard laughed, and that laugh was his last for many a long day. When the men were gone, Eveleen crept out from behind a large chest where she had been hiding, and her eyes were red and swollen, as she went over and threw her arms around her father's neck, and drew him down to a seat near the fire. "Don't cry, father dear," said the affectionate child, seeing her father's cheek wet with tears, "don't cry—I can't bear to see you cryin'."

About an hour after the departure of the Scripture-readers, while the O'Dalys were still talking

ever the shameless conduct of the proselytizers, all
of a sudden

"There was heard a heavy souul, as of arm'd men the tread"

On it came, nearer and nearer, until it stopped
in front of the house; then there was a clang, as
of arms grounded, and the girls looked at their
father in speechless terror. The old man was pale
as death, and his lips were closely compressed; he
tried to stand up, but his trembling limbs refused
to perform their office, and he sank again into his
chair, and looked piteously around on the three
terror-stricken creatures whom he had no longer
the strength to protect. "So they're comin' at
last, children!" he said, in a smothered voice:
"They're as good as their word, the black-hearted
villains. May the Lord grant us patience, an'
strenth to bear what he's layin' on us!"

"Oh father! father dear! what's to become of
you, at all?" cried the now weeping girls, as they
wrung their hands in despair.

"Shame! shame, children!" said their father.
"Will you fly in the face of God?—dry up your
tears, an' keep quiet now, for the love of God, an
don't let them vagabones see you cryin'. Don't
give them that satisfaction."

By this time Ousely's bailiff and two of the policemen were in the kitchen, and having read the
process of ejectment, commanded Bernard O'Daly
to quit the premises forthwith.

" Well ! God's will be done !" ejaculated Bernard. "I'm a long time in it now, an' so were my father an' my grandfather before me, an' it's little any of us thought that the day would come when we'd be turned out of it. Be quiet there, children !" he said, with calm dignity. " Not a word with you. Maybe," said he to the bailiff, who was one of the Jumpers, " maybe you'd allow us to take a couple of quilts an' a blanket or two."

" No, nor the devil a stitch, old fellow !" returned the bailiff, who, with the policemen, was already gathering the movable furniture together. " Be off as fast as your legs can carry you. Stir yourselves, lads !" Kathleen went over to where her father's old overcoat was hanging on the wall, and would have taken it down, but Sweeney, the bailiff, stopped her with a brutal execration. " Leave it there, and be d——d to you !"

" Oh ! Mr. Sweeney, dear !" cried the heartstricken girl, " won't you let me take my father's coat —God help him ! he hasn't much on him now, to keep out the winter's cold, and it will be the death of him to go out such a day as this without his overcoat, and God knows where he'll get a shelter !"

" I don't care a damn whether it does or not !" returned the heartless ruffian. " My orders are to seize every thing that can be sold. Out with you now, the whole set of you—do you want us to have the trouble of lifting you out? Come here, then, Stephens, and you, Tomkins, we'll give these pretty girls a lift, since they don't choose to use their delicate feet!"

"For the love of God come away, father!" cried Bridget, taking hold of her father's arm. Kathleen pressed close on the other side, but still the old man lingered. He looked wistfully at the old straw chair, wherein Honora used to sit, and he was sorely tempted to ask for it, but he knew well what the answer would have been, and kept in the words which rose from his heart. Still he stood a moment with his eyes fixed on that dear old chair, and as he gazed, the tears, before pent in, came slowly forth, and chased each other down his cheek. His daughters well understood his feelings, and shared his mournful thoughts, but no one spoke, until Sweeney, seeing them linger a moment, came behind, and gave Bernard a push that sent the grief-worn old man some yards outside the door, where he would have fallen, had not one of the policemen caught him. At this moment, there was heard a loud noise in the rear of the house, and the word went round amongst the policemen, "It's young O'Daly, and there's a whole crowd of ragamuffins with him!"

"Stand close to your arms, men!" cried the chief constable; "we may have work to do!"

Owen and his friends had gone to the back door

seeing from a distance that the front was well guarded, but they found the door barricaded against them, and then Owen sprang over the gate at the end of the house, (charging the others not to follow him,) and was making up to his father and sisters, within the ring, when the chief constable laid his hand on his shoulder:

"Stand back, young man, stand back! you cannot pass here!"

"But my father and my sisters are there—may I not speak to them?"

"Yes! but not here—let them pass out, men!"

But just then Kathleen discovered that Eveleen was not with them, and she was just on the point of calling to her, when the little girl was seen through the open door struggling in the hands of Sweeney and the policemen within.

"Eveleen!" cried her father—"Eveleen! my child! my child!"

Owen waited for no more, but dashed through the ranks of the policemen, putting aside with his hand their bristling bayonets, and before any one could prevent him, he caught Eveleen in his arms, and was already outside the door, when Sweeney called out "Stop them there—don't let them pass, I say! The little witch has been picking up things in the room. Search her, captain!"

"Stand back!" cried Owen, in a voice that startled all within hearing; "stand back there

captain, or whatever you are. Don't dare to lay
a finger on the child, or—will you dare?" he
shouted, raising the screaming child with one arm,
while with the other he grasped at the officer's
throat. "Back now, and let us pass, or I'll choke
you—aye! if there was fifty of your bayonets
about me. Ha! ha!" he laughed, or rather
shrieked, as the amazed chief made a sign for his
men to make way. "Ha! ha! you're a wise man,
I see!—you know it's not safe to play with a mad-
man—he doesn't regard bayonets! Come on
now, father!—They'll not ask to stop us!"—he
added bitterly, as the stupefied old man followed
close behind, almost carried by his two elder
daughters. By this time, the men from behind
the house had got around to the front, and a formidable aspect they did present, for they were all
armed with spades. On seeing the miserable
group of which poor Bernard was the centre, with
Eveleen clinging to his arm, the men became almost frantic with fury.

"Ah, then, Owen," said one stout, burly fellow,
no other than Patsey, our old acquaintance; "Ah,
then, Owen, how can you stan' that? By the laws,
I'm willin' to lose the last dhrop o' my blood—an'
begorra I will, too, if it's a wantin'. Boys!" said
he, addressing his companions; "is it come to that
with us, that we'd stan' by an' see Barney O'Daly

an' his family turned out on the world on a cowld winther's day?"

"Faith an' it's not, then," cried Brian Han ratty, making a flourish with his formidable weapon.

"Let us give it to them now, once for all," cried one; and another shouted,

"Many a good turn we owe them."

"Look at that divil's bird, Sweeney, the Jump er!" cried a third; and so great became the up roar of angry voices, that neither Bernard nor his daughters could make themselves heard. The policemen began to put themselves on the defen sive, and as the crowd of angry assailants was every moment increasing, the affair was becoming serious. For some minutes there was nothing heard save the deep voice of the police officer, as he formed his men into a square, and the fierce threatenings of the surrounding crowd, now swelled into a multitude. In vain did Owen O'Daly try to persuade his father and sisters to retire to some of the neighboring houses.

"No, no, Owen!" said his father, " we'll not stir a step till you're with us. If we went, God only knows what might happen. But come you with us, *acushla!* an' we'll go any where—any where out o' this!"

Here there was an interruption, owing to the arrival of Mr. Ousely, who rode up at full speed, and dashed in amongst the crowd, amid a volley of fearful execrations.

"There he goes, the tyrant!—make way for him there, or he'll tramp us all down—an' that same 'id be bread an' butther to him—ah! you hard hearted villain! your own hearth-stone 'ill be as cowld as Bernard O'Daly's some o' these days, an' there 'ill be no one sorry for you, Ousely! when *your* time comes!"

Ousely made no answer, but kept turning from side to side with a scowl of fierce defiance. Having spoken a few words with the police officers, he commanded the people to disperse, or otherwise he would read the riot act, and order the police to do their duty.

"No, nor the sorra foot we'll stir out o' this!" cried one and another. "Come on, boys, as we couldn't get any work this mornin', we'll give a hand here!"

Bernard laid his hand on Owen's shoulder, and begged him, for God's sake, and his sake, not to raise a hand against any one.

"What good can you do us, Owen dear? you can't put us into the house again, for it's not ours any longer, an' you'll only be the cause of blood shed, an' maybe loss of life. Go, Owen dear! for I'm not able, an' persuade the poor fel.ows to scatter peaceably afore the Act is read. Do, *astore machree!* an' you'll have your father's

blessin'—if there was a life lost on my account, it
would break my heart—it would indeed!—go!—
go!—or you'll be too late!"

The fiery youth could ill brook such a mission,
but he had never disobeyed his father, and would
not begin now, when his heart was crushed beneath
a double load of sorrow. Going over to the most
violent of the men, he begged of them to desist,
telling them what his father had said. There is no
knowing what effect the message might have had,
but just then there was a cry of "The priest! the
priest!" and a ready passage was opened for him
as he rode up, followed by Phil Maguire on his
white pony.

"Where are they?" said Father O'Driscoll,
after exchanging a cold salute with Ousely;
"where are Bernard and the children?"

"Here they are, your reverence!" said a score of
voices; "here's poor Bernard sittin' on the cowld
stones."

"Yis, here we are, Father O'Driscoll!" said
Bernard, his tears breaking forth anew. "Here's
myself an' the girls, an' poor Owen, without a roof
to cover us, blessed be God for it—it's *his* will, or
it wouldn't happen to us."

"May the Lord comfort you!" said Father
O'Driscoll, as he alighted from his horse, and
squeezed the old man's hand. "But don't despair,

Bernard! God's arm is not shortened, and He sees what is going on!"

"Will you get out o' the way, an' bad manners to you?" cried Phil Maguire from behind. "Beggin' your pardon, Father O'Driscoll! I didn't mane you!"

"Hurra, boys! clear the way for Phil Maguire—himself and Bernard's old friends. God bless you, Phil! every day you rise!"

"Thank you, boys, thank you kindly!" He had now made his way up to Bernard, and taking hold of his hand, he shook it as though he would shake it off him, and looked him straight in the face, but said nothing. When he came to the girls, he did the same, and when they were breaking out into ejaculations of sorrow, he stopped them short, with "Never mind, girls, never mind! all's not lost that's in danger! Here, Owen, stay with them—I've a word to spake to Misther Ousely there!"

By this time the priest had laid his commands on the people to disperse quietly, but every one was anxious to hear what honest Phil had to say to the landlord, and there was a dead silence.

"Misther Ousely!" said Phil, touching his hat according to custom; "Misther Ousely! a word with you, sir! Wouldn't you take my note for what arrears is on the O'Daly farm—it's only sixty five pound, an' you know I'm good for more than that! I never gave a note before to any one, but I'm willin' to do that an' more, sooner than see Bernard O'Daly put out of his place!"

"I'll have nothing to do with your note," said Ousely, in his surliest tone. "If you'll hand out the money, I'll let them stay—not otherwise!"

"Because you know well enough," said Phil, "that if I had the ready money, I wouldn't offer you a note. But no matter for that—take the note—" and he held it out to him—"I'll forgive you all, if you'll only do this! Do, for God's sake, Misther Ousely!"

"Not for any sake!" said Ousely, fiercely, "O'Daly has brought it all on himself, and the law shall take its course. As for you, Maguire, you had a different tune some weeks ago, when I sent for you—do you remember that?"

"I do," returned Phil stoutly, "an' I thank God I'm jist as unbehouldin' to you now as I was then. God pity them that *is* in your power this blessed day. Boys!" said he, turning abruptly to the listening crowd; "boys, I'd have you all to know that if poor Bernard O'Daly is sittin' there with his children, without a roof to cover them, it's because neither he nor his 'id have anything to do with the Jumpers. It's because he wouldn't turn his back on his religion, an' make a god of the soup-boiler or the stirabout-pot. That's the thrue raison of his boin' turned out—it's not the rent. at all."

A yell of execration followed, and the excite
ment became so great, that Ousely was glad to
dismount and take shelter in the house.   The police
themselves were evidently alarmed, and drew
close together with bayonets pointed, waiting for
the attack ; they had not room to take aim, being
closely hemmed in by the laborers, with their
fearful looking weapons raised aloft, ready to
wreak vengeance on those who had been so often
the instruments of tyrannical oppression.   Kath
leen and Bridget O'Daly covered their eyes with
their hands, and begged of their father and Owen
to leave that terrible scene, but no one listened to
them, nor to Eveleen, though she kept screaming
and clapping her hands.   Already were the spades
uplifted for a crushing blow, and the pale faces and
compressed lips of the policemen, as they grasped
their bayonets, showed them prepared for a mortal struggle.   Not a word was spoken on either
side, for the passions of all had settled down into
the fearful calm of desperation, and it seemed that
no earthly power could restrain the tide of destruction, but suddenly the voice of Father
O'Driscoll was heard : " I command you to fall
back," said he, " and to shed no blood !   In the
name of God, do what I bid you !"   There was
heard a low murmur, like the subterraneous growl-
ing of pent-up elements, and then the crowd fell
back, and the spades were lowered, and the

policemen began to breathe more freely, and even
Ousely put his head out of the door-way. At his
appearance, the storm was well-nigh raised again ;
there was a cry of " Don't let Ousely escape !
Now's the time to pay him for all !"

" Silence !" cried Father O'Driscoll, " Silence ! —
not a word more, I charge you ! If the man has
done wrong, leave him to God—he is the Aven-
ger—not you !—The first thing *we* have to do is to
seek a shelter for this afflicted family."

" 'Deed, an' that won't take you long, Father
O'Driscoll," said Phil Maguire, briskly. " They'll
not want a shelter while I have one to give them.
There's room enough for them in my place above.
an' they're as welcome as the flowers in May !"

" The Lord bless you, Phil !" said Bernard, fervently. " It's you that's always the thrue friend."

" Yes, Bernard !" said the priest, " the friend in
need is the friend indeed ! May God bless you,
Phil !" It was all he could say, but the warm
grasp of his hand did Phil's heart good, for it assured him of his fullest approbation.

" Come, now," says Phil, beginning to bustle
about in his old way, " what will we do for a cart,
to take Bernard an' the girls up—for I know they're
not able to walk ?"

" Hurra !" shouted those who were on the out
skirts of the crowd ; " hurra ! hurra ! for granny

Mulligan! Long life to you, granny! It's you that's always to be had when you're a wantin'."

"Clear the way there!" cried others. "Here's granny Mulligan herself, with a cart—more power to your elbo v, granny!—that's it, granny!"

Sure enough, it was granny Mulligan herself, who now drove up, standing in Phil Maguire's cart, and managing the reins with as much vigor as though she were but "sweet five-and-twenty," as she said herself.

Heedless of the warm gratulations of the numerous by-standers, granny drove right up to the O'Dalys, and there stopped. Not a word could she speak for a full minute, during which time she cleared her throat more than once, but at last she found voice to speak:

"So they put you out," said she; "out of the ould walls where your forebearers lived in pace an' plenty—an' it's all for religion—religion, inagh!" she repeated, with ineffable scorn—"sure, isn't the three known by its fruit, an is'nt sich a sight as this enough to shew what *their* religion is—the curse o' God villains—a decent body ought to wipe their mouth afore they'd mention them or their sham religion!—ah! you're in there, Ouse-ly!"—she just then caught a glimpse cf him through the window—"an' there's your right-hand man, Alick Sweeney—the white-livered dog!—ah! there'll come a day for all this—mind my words

but there will !" and the excited old woman shook
her fist at the squire, from her elevated position,
to the great amusement of the spectators, policemen and all.

"Blood alive, granny ! how did you know we
wanted the cart ?" cried Phil—"or was there no
one else to come with it ?"

"The sorra *that* there was, Phil," returned
granny—"an' bedad, myself and Nanny thought I
had best get in an' dhrive myself, so atween us
we tackled the horse, and off I set, an' it's well I
did, too.   Get in here, girls—ah, then, Eveleen,
my poor, fair-haired *colleen*—is it come to this
with you ?   Bernard ! poor man ! get in here—
Owen or Phil will dhrive back, an' I'll walk !"
So saying, she motioned to Owen to help her to
alight, but Phil interposed, and made her stay
where she was.

"No, no, granny ! stay where you are—we'll
walk beside the cart !"

"Well, make haste, then, all o' you, for Nanny
has a fine dinner ready, an' it'll be spoiled if you
don't hurry."

Father O'Driscoll now came forward with a
smiling countenance, and extended his hand to
granny :

"So you tackled the horse and drove down
yourself, granny ?" said he.   "You really deserve
credit.   I did not think you had been so active !"

"Is it me, your reverence!—oh, then, indeed, I
could do more nor that if I was put upon. The
like o' this makes an ould body young again!—ah!
you villain!" she cried, shaking her fist again at
Ousely, who just then appeared at the door—
"you hard-hearted villain! it'll come down on you
hot an' heavy, so it will!"

"My good woman!" interposed the chief of
police; "I cannot allow you to talk so to Mr.
Ousely!—I'm here to keep the peace."

"*You* keep the pace!" cried Brian Hanratty,
who stood near him. "Is it you keep the pace?—
why, bad manners to you for a spalpeen, wasn't it
Father O'Driscoll that kept the pace—if *he* wasn't
here, I'd like to know where *you'd* be by this time!
Three cheers, boys, for his reverence." In an
instant every caubeen was in motion, and cheer
after cheer rang out through the grey misty sky,
awaking the echoes of the neighboring mountains.
It was a cheer that Connemara well knew, for there
is none other that comes so directly from the heart
in that wild and remote region, as that which re-
sponds to the word "*soggarth!*"

"And now a groan for Ousely an' the Jumpers!"

"An' a groan for the lyin' Prodestan' bishop,"
cried another—"him that said there was ten thousand Jumpers in Connemara! Faix, if we had

him here, we'd put the lie down his throat, an' a
bouncer of a lie it is, too !"

The groan that followed was more than a groan—
it was a yell of fierce defiance, and it was renewed
again and again until the small party of policemen
began to quake once more.   But they had no need,
for their guardian angel was still present in the
person of Father O'Driscoll.

In a few minutes after, the procession moved
away, and a tumultuous one it was, too, for every
man there seemed to have made up his mind to
accompany the cart, by way of forming a guard
of honor.   It was a strange sight—a truly Irish
sight—to see that grey-haired old man, with his
three daughters and his young son, turned out of
the house where they had all drawn their first
breath, and their fathers before them for generations back—the house which had been improved
and made comfortable by their ceaseless industry
to see them turned adrift on the wide world with
out a penny in their pockets, just at the opening of
winter.   And then to see the numerous escort, all
vieing with each other in paying attention to the
poor homeless family—all eager to do them any
little office of kindness which their own poverty
would permit them to offer—the rough man of
labor, softened to woman's tenderness, and for-
geting his own half-starved condition in his keen
sympathy for the O'Dalys—for them who had

often relieved him in by-gone years. The whole scene was one of heart-rending interest.

Father O'Driscoll rode beside the cart, alternately consoling Bernard, and soothing Owen's exasperated mind. At Phil's request, he accompanied them to his house, and partook of Nanny's "fine dinner." When they reached the foot of the lane, the crowd separated, having first given three cheers for Phil Maguire, and three more for Bernard O'Daly; then again for Father O'Driscoll, who gave them his blessing at parting, and warmly thanked them for their prompt and cheerful obedience. "Long life to you, Father O'Driscoll! it 'id be a bad day for us if we didn't obey you!" said one; "I hope that day 'ill never come!" said another, as they turned away, each taking the road to his own desolate cabin.

Owen was moody and silent all the evening, notwithstanding the persevering efforts of his friends to divert him from his gloomy thoughts.

"But tell me, Eveleen dear!" said Kathleen, suddenly, "what was it that kept you after us?"

"Why, sure I wanted to get something that had belonged to poor mother. And when I heard the ugly, wicked man saying that you mustn't take anything, I thought I'd steal into the room, and—"

"And what did you get, dear?" said her father.

The little girl put her hand in her pocket, without speaking, and pulled out a pair of large old-fashioned beads, which were at once recognized as having been her mother's. This sight drew tears from all present--even Father O'Driscoll's eyes were moist.

"An' so," said Bernard, "it was to make sure of the beads that you staid behind, Eveleen?"

"I wanted to get mother's specs, too, Father," said Eveleen, quietly; "but the men came on me before I could find them, an' they wanted right or wrong to see what I had in my pocket, before they'd let me go. That's what made me cry out the way I did, for I was afraid that they'd take the beads from me."

"Poor child!" said Father O'Driscoll with a melancholy smile, "you might not have been afraid of that—they do not covet such things, unless to throw them in the fire, or some such thing."

Granny Mulligan was installed in the chimney corner for that evening, and if she had been a queen, she could not have been treated with greater respect. She was in all respects the queen of the feast, and a gay old queen she was.

CHAPTER XII.

I mean to show things as they really are,
Not as they *ought* to be, for I avow
That till we see what's what in fact, we're far
From much improvement.—BYRON's *Don Juan.*

WHEN we last saw Eleanor Ousely, she was going with her mother, to call on Mrs. Hampton, the wife of Captain Hampton of the 27th. Mrs. Hampton was an Englishwoman of limited education, and full of strong prejudice against "Ireland and the Irish." Still, this was more the effect of an erroneous system of training, than of any natural antipathy to the Irish or any other people, for, on the whole, Mrs. Hampton was a good-natured, well-meaning woman, ready and willing 'to do a good turn whenever it was required. When Mrs. Ousely and her daughter had set down Mr. Ousely at the court-house, they drove to Mrs. Hampton's, and were shown into the drawing-room, where they found Capt. Hampton, with one of his subordinates, a foppish-looking young gentleman, who was introduced as Lieutenant Gray. Mrs. Hampton insisted that the carriage should be driven into the yard, " for," said she, " you must wait for our luncheon—it is a long time since you spent an hour with me, so now you shall spend *two*, at least."

"But Mr. Ousely may be waiting for us," said Mrs. Ousely, in her quiet way.

"Well! let him wait!" returned Mrs. Hampton, quickly. "Surely you're not afraid of him— it is only the Papists who hold him in awe, I fancy!"

"Caroline!" said her husband, in a significant tone, and then he gently turned the conversation into another channel. In this he was assisted by Eleanor, who well knew that Mrs. Hampton was sometimes " more candid than polite."

"How do you like Connemara, Captain Hampton?" said Eleanor; " it is a very wild region—is it not?"

"For what I have seen of it, Miss Ousely," replied the Captain, " I like it very well. Somewhat different, indeed, from what I had expected to find it, but that is nothing strange, for I might say the same of almost every place where I have been, in Ireland. It is surprising how little we English know *at home* of the actual condition of Ireland, or even of its scenery. It is a very beautiful country!"

"Beautiful, indeed!" said Mrs. Hampton, contemptuously; " I should like to know what you call beautiful!"

"Well, my dear!" said the captain with a smile,

" better judges than either you or I have long ago
given that decision, and I believe it passes current
every where. What a pity, Miss Ousely! that
such a country should be inhabited by a race of
paupers! The poverty of Ireland is so great, so
lasting, and so general, that one is almost tempted
to think that a curse hangs over this fair and fertile
land !"

" And so there is, Frederick !" said his wife,
earnestly ; " the whole world knows that there is
a curse on Ireland—the heavy curse of Popery."

Eleanor and Hampton exchanged a meaning
glance, and even Mrs. Ousely smiled. " That
naughty Popery has much to answer for, my dear
Mrs. Hampton !" said Eleanor, " if it be the cause
of all the misery which exists and has for ages
existed, in Ireland. But surely," she added with
an arch smile, " surely, we may hope for a speedy
improvement—Popery, you know, will soon be
banished from Ireland, and then all will go on well
—we shall have the millennium, as a matter of
course."

" Yes, but who's going to banish Popery ?"
observed Mrs. Hampton, who did not well understand irony, and, therefore, took Eleanor's words
in their literal signification. " I'm quite sure that
the missions here are not making much real
progress, though they make a great fuss about
what they do !"

"Why, you forget, my dear Caroline," said her husband, gravely, "that his grace of Tuam—I mean our own dignitary—has publicly boasted of having ten thousand converts in his arch-diocese. Recollect yourself, my dear!"

"Well! of course he knows best," said Mrs. Hampton, "but if any one else said it, I should certainly set it down as a wholesale mistake, or something else. But, of course, archbishops never lie. I only hope that the converts are of a more reputable character in other p'aces than they are here."

"As to that," said the Captain, "I suppose—nay, I believe they are pretty much the same all through."

"For my part," said the sub, in a soft, lisping tone, "I never trouble myself much about such things, but I must say it is *rather* hard that we should be obliged to escort these wretched converts to church, as is the case in many places. I have been several times obliged to do it, and really I did feel exceedingly small on those occasions!"

"And the worst of it is," said Hampton, with his soldierly frankness, "that the precious *converts* might have gone to church, in every one of those instances, without our company. They merely represented themselves as being in danger, in order to make themselves of some importance. In fact, our being with them often drew ridicule and insult

upon them, that I am confident they woul
otherwise have escaped."

"Yes !" said Gray, "I remember, one wet Sun.
day, when we were stationed in some out-of-the-way
place down near Achill, we had to guard a half
dozen or so of these stirabout converts to church,
and, 'pon honor, I got my best over-coat completely
spoiled—I positively did."

" Why, how did that happen, Mr Gray ?" demanded Eleanor, trying to keep from laughing.
" Were *you* mistaken for a convert ?"

" No, Miss Ousely," lisped the Lieutenant, with
an air of offended pride ; " no, not quite so bad as
that, but the people began to quiz the confounded
converts about having a guard of honor, as they
called it, and the others answered back, doubtless
depending on *our* protection, whereupon there was
some mud thrown at *them*, but unluckily I got part
of it. I really could have seen the converts far
enough at the time—in fact, any where but where
they were."

" But why blame the converts, Mr. Gray ?—
surely it was not their fault?"

" Why, not exactly—though it was, indirectly at
least—but the fellows who threw the mud were so
sorry, and made so many apologies for hitting me,
that I could not bring myself to be angry with
them, poor devils—I must say I respected *them*
more than I did the converts—so-called."

"Well, really," said Mrs. Hampton, "though should be glad to see Popery abolished, yet somehow it don't seem as if there's any great chance of its being so in our time, and I must say that these ten thousand converts—dear me! I hope the archbishop didn't make a mistake—are not worth all they cost—what with the soup and stirabout, and never-ending collections taken up for them, and the guarding them to church, and I don't know what all."

"What with one thing and another, Caroline," interrupted her husband, "we might say to them, collectively, as Cora said to her child: 'Thou art dear bought!' Poor Gray is ready to endorse *that* opinion. But what about the lunch, Carry—I thought you promised us some?"

"And so I did, Frederick—and I was really forgetting all about it!" She then rang the bell, and ordered luncheon to be served in the breakfast-parlor, whereupon the captain offered his arm to Mrs. Ousely, and the lieutenant was so eager to secure Eleanor for the journey down stairs, that he came near stumbling over an ottoman which lay between them.

About noon, the carriage was ordered round, and the ladies proceeded to pay their remaining visits, having obtained a promise from the Hamptons and Lieutenant Gray to dine at the Hall on the following day. On reaching home, Mrs Ousely asked John, who opened the door, whether his master had got back yet.

"Why, Lord bless you! no, ma'am!" replied John—"sure he's down at Bernard O'Daly's."

"At Bernard O'Daly's!" repeated both ladies in surprise ; "what in the world is he doing there ?"

"So you didn't hear what happened then? Sure the O'Dalys are ejected—turned out root and branch—and there was near bein' bad work there—only for the priest, they tell me, the master, and the police, and all would have got something to remember while *they* live—and maybe it's kilt they 'd have been all out, for sure all the laborers ran out from the town with young O'Daly, when word was brought to him of what was goin' on ; and besides, the people gathered from far and near when the word went out, and they say there wasn't sich a gatherin' seen this many a day."

Both mother and daughter stood aghast on hearing this, and for a moment neither could speak.

"But are the O'Dalys left without a roof to cover them ?" said Eleanor at length.

"Hut, tut, Miss Eleanor! don't you know very well," said John, "that the likes o' them wouldn't be long on the road—no thanks to them that took the shelter from them !—why they weren't many minutes out o' the house, when Phil Maguire was there—and they tell me he offered to pay the whole arrears, but his note wouldn't be taken—

and didn't granny Mulligan—the owld beggar
woman—drive down Phil Maguire's cart, and they
were all taken up *there*.   Oh, by the laws, Miss
Eleanor, the counthry would be gone to the dogs
altogether, if *they* 'd be left without on the road—
though many a dacent ould family *is*, God knows !
Howandever, you may thank Father O'Driscoll,
or there'd be black sorrow for miles round this
blessed day, and—well ! no matter !—it's best as
it is—and thank God that *you* haven't got the sore
heart, let who will have it !"

It was late in the evening when Ousely came
home, and even then his manner was still flurried,
and his face paler than usual, from the effects of
the recent agitation.   During dinner he spoke little,
and what he did say was cold and stern, without
any allusion to what had passed.   He asked where
the ladies had been, and they answered in few
words.   On the whole, the meal was any thing but
pleasant, for there was a gloom hanging over all,
and the very *viands* on the table seemed to have
lost their usual flavor.   At an early hour—much
earlier than usual, Eleanor retired, and the others
soon after followed her example.

On the following day, Ousely seemed to have
recovered his usual spirits, and undertook to give
an account of the proceedings of the previous day
Eleanor and her mother listened with apparent
somposure, and made little or no comment, Mrs

Ousely never daring to find fault with her hus
band's conduct, and Eleanor well knowing that
there was no good then to be effected by her interference. She was sick at heart, and felt as though
she would have given worlds to be anywhere but
where sne was. The almost daily recurrence of
these scenes was a source of unmitigated torture
to her sensitive mind, and each tragedy, as it occurred, seemed to weaken her affection for her
father, who was the author and executor of
them all.

Her tenderest sympathies were with the poor,
suffering people, who were made to endure such
unheard of miseries, and who bore them with such
unprecedented patience and resignation. Their
sufferings, and their virtues, and their humble piety
were constantly in her mind, and these, coupled
with her acquired knowledge of the Catholic religion, and her conviction of its divine origin, gradually brought her mind to a fixed and steady
resolution to cross the Rubicon, and take refuge in
the land of peace. But for the present she kept
her decision to herself, awaiting a more favorable
opportunity to disclose it, even to her mother.

In the course of the day, Mr. Ousely told his
wife that he was going to give O'Daly's place to
Alick Sweeny. "The fellow deserves something,"
said ne, "for he has done me good service, and

besides he's a convert, and I want to encourage him. It will incite others to follow his example."

"Well, my dear, you know best!" was the meek rejoinder of Mrs. Ousely, but not so Eleanor.

"My dear father!" said she, "you cannot but know that there is not in the whole country a more disreputable person than that Sweeny. Why, his name was a by-word long before his conversion— if conversion you choose to call it, and we have not heard that he is anything improved of late. Surely you will not think of giving *him* that fine farm and farm-house, on which the O'Dalys have expended so much money. Just think of how it will look, father—think of the man's character!"

"Why, what the d—l, Eleanor! can't a man do what he likes with his own, without being called to an account for it? I tell you that Sweeny *shall* have the place, so there's no use in talking any more about it. If that obstinate old fool, O'Daly, hadn't been so stiff, he might have been in it still. It was only yesterday morning that I sent to offer him terms, but he wou.dn't hear a word the men had to say!"

"And who were *the men*, father?" asked Eleanor in a careless tone.

"Why, Hanlon and McGilligan—who else?"

Eleanor smiled, but said no more. She had heard all she wanted to hear, and she thought it best to take another opportunity of reasoning with her father on the disgraceful project of putting the despicable Sweeny in the ancient holding of the O'Dalys. She and her mother persuaded Mr. Ousely to ride over to Clareview early in the morning, and engage the Dixon family for dinner.

"By Jove I will!" said Ousely, "and we'll have capital fun, for I know Hampton is a d——d good fellow, and so is Dixon, though he *does* keep company with the priest; and then that young Trelawney is a devilish fine fellow, though not the best hand in the world ' to push about the jorum' —O'Hagarty must come, too, by ——, for, like old King Cole, he's a merry old soul, and a merry old soul *is* he! Hillo, Ben! bring out Tom Turpin (his favorite roadster). I'll be off at once!"

In due time for dinner came Captain and Mrs. Hampton and Lieutenant Gray, Sir James Trelawney, Mr. and Miss Dixon, but, to the great disappointment of Mr. Ousely, the Reverend Bernard did not make his appearance, though dinner was kept back a full half hour.

"For whom are you waiting, Ousely?" said Mr. Dixon, seeing that his host kept watching the door.

"For the Reverend Mr. O'Hagarty!" returned Ousely. "He promised to come without fail."

"Humph!" said Mr. Dixon, "I rather think you needn't wait any longer. Eh, Sir James?"

Trelawney shook his head and smiled.

Hampton laughed. " Why, really, Mr. Dixon !" said he, " I think ' more is meant than' meets the ear' in your remark "

" 'Pon my honor I think so too," said Gray.

Dixon kept looking from one to the other with a provoking smile. At last he turned to his daughter, who had been telling Eleanor something in a low voice, that made the latter burst out laughing.

" Shall I tell, Amelia ?"

" Just as you please, father. I have been making Eleanor as wise as myself."

" Why, hang it, Dixon, let us hear it, whatever it is !" cried Ousely, with a gesture of impatience.

" Take your time, Ousely," rejoined the other, " ill news comes soon enough, and I think you will all agree that this is bad news. As we came by Alick Sweeny's on our way hither, our ears were assailed by some unusual sounds, and looking in, we perceived the Reverend Bernard. minus his coat, his stolid countenance flaming red, and Alick Sweeny belaboring him, might and main, with a stout shillelagh. His reverence was evidently the worse for liquor, in other words, gloriously drunk, and if he didn't bawl, and jump, and cut capers through the floor, no man ever did. And the fun of it was, that Sweeny, the rascal ! was just as cool as a cucumber, and kept saving at every

blow: 'There's for you, now! take that now!' will
you do it again, you beast?' with sundry other
compliments of a like character."

Every one present laughed aloud, except Ousely,
who seemed far more inclined to cry, and the sight
of his doleful countenance made the others laugh
still more.

"Why, d—n the villain—I mean that Sweeny!"
said he, after a short pause—"what did he do that
for?"

"For a very good reason," replied Dixon,
coolly; "because the fellow had been making love
to his wife in his absence, and went about it so
roughly that the gentle dame complained to her
husband, who returned thanks for his attention in
the way I have described."

"Still the scoundrel had no business to go so
far!" cried Ousely. "His wife isn't always so
squeamish, and he knows that well enough. I'll
be hanged if I'm not even with him for that—he
may go whistle for a farm now."

Here dinner was announced, and the gentlemen
proceeded to "take the ladies" in the order pointed out by Mrs. Ousely. Sir James anticipated
the word of command, by drawing Eleanor's arm
through his own, whereupon the Lieutenant made
up to Amelia with his best bow. As they went
down stairs, Eleanor said to Trelawney :—"I wish
you had been at Captain Hampton's yesterday

when we were there. I was very much amused
by certain reminiscences of the captain and lieutenant Gray concerning the proselytizing system.
I must bring the subject round again, for your special benefit."

"You are very kind," said Trelawney, "to
think of me in any case." After a moment's pause,
he added : "I, too, saw something of interest yesterday. Have you been to Jenkinson's school
ately ?"

"No," said Eleanor; "not since I was there
with you."

"Well! I was there yesterday, and what do you
think they have got, by way of improvement ?"

"I am sure I cannot tell."

"Neither less nor more than a huge trough,
similar to that used for swine, for the greater fa-
cility of administering the stirabout."

"Why, surely, you are not serious, Sir James ?
You don't mean to say that they make the children
eat from a trough ?"

"Precisely so," replied the baronet—"I mean
just what I say. The thing was exhibited to me as
a capital contrivance. Oh, blessed effects of the
New Reformation !" he added, bitterly. "Re-
ducing the children of the poor to the level of the
brute creation, and yet for this they are to barter
the faith of their fathers—the old, venerable faith
that raised them above the wants and woes of

arth !"   As he spoke thus, with unusual earnestness, he felt a slight pressure on his arm, and met
Eleanor's dark eyes raised to his for a moment
with an expression that made him thrill all over,
for there was in it both sympathy and approbation.
No more was said at that time, for just then they
reached the door of the dining-room, but all that
evening Trelawney felt happier and more hopeful
than he had for a long time past.

The evening wore away rapidly. "Laugh, and
song. and sparkling jest went round," and the gentlemen lingered long over their wine, so that it
was fully eight o'clock when they joined the ladies
in the drawing-room. The company had formed
itself into small detached groups of two and three
here and there through the spacious apartment,
and Amelia had just taken her place at the harp,
when a servant came in to tell Mr. Ousely that
there was a person below stairs who wanted to see
him.

"Do you know who it is, Billy ?"

"Faith, an' I do, sir. It's Misther O'Hagarty—
the priest that was, sir. Between ourselves, your
honor," lowering his voice to a confidential tone ;
"Between ourselves, he's not the soberest in the
world. He's as full as a piper !"

"What the d—l brings him here, then ?" cried
Ousely aloud. "Tell him I can't see him now. '

"I did tell him that, sir, an' he was near sthrikin
24*

me. He says he must see you, let what will come
or go!—you may as well come at once, your
honor, for he'll not go without seein' you."

"Confound him for a beast!" growled Ousely,
as he rose to follow the servant.

"Fie! fie! Ousely!" cried Dixon, from the
other side of the room—"Is it thus you speak of
a pillar of the New Reformation—a valued *protegé*
of the Priests' Protection Society? Go and see
him by all means, lest he should be tempted to
come up here, an honor which none of us covets,
I am sure! He must be *non compos mentis*, by this
time, I think!"

Ousely went down with visible reluctance,
whereupon the company began to discuss the subject of the proselytizing system, and it was generally admitted to be one of the grand humbugs of
the age.

"And a humbug which is likely to produce the
most serious and lasting evils," said Dixon—"that
is, as far as it produces any thing. Now, I am a
Protestant. I belong to the church by law established in these realms, nor have I the slightest in
tention of ever leaving it, for to tell the truth, I
neither know nor want to know, any other form of
Christianity, but I am perfectly convinced, and
that from ocular demonstration, that there is not
the shadow of a chance of effecting a change in the
religion of the Irish people. The Catholic religion

is a part of their very nature—it is intertwined with all their dearest and most glorious associations; it is peculiarly adapted to the nature of man; it is essentially a religion of comfort and consolation, and, therefore, dear to the suffering and the poor, and the consequence is that it is scarcely ever rooted out from a country where it has once been planted."

"Witness our own England!" said Hampton, "where it is now springing up with renovated strength, after an interval of three hundred years, during which it was supposed to be dead!"

"Oh! it was only taking a nap!" said Amelia. "Its slumbers were watched over all the time by those venerable worthies, the Vicars Apostolic!"

"But, talking of the Church of Rome here in Ireland," resumed Hampton, "I can well understand many of the reasons why all attempts at Proselytism should prove abortive. Now, let us take it as our starting-point, that salvation is certainly to be attained within the pale of the Roman Church—though none of us will approve of her appropriating it exclusively to herself—then, let us remember the long series of ages during which it has flourished in this country—let us consider the almost innumerable multitude of saints and heroes, poets and sages, whose names are held in fond remembrance by the Catholics of Ireland; in fact, there is scarcely a name which they hold dear or sacred, that is not intertwined with Catholic associations—nay, identified with Catholicity itself. Look at their O'Neills and O'Donnells, and, indeed, all their warrior-princes; were they not fighting the battles of their religion as well as of their country—and on that very account it is that their names and their actions are enshrined in the hearts of a grateful and a religious nation. Look over this island, from east to west and from north to south, and you will see it covered, literally covered with monuments of Catholic piety and Catholic worship. You will see monasteries, and cathedrals, and churches, and stone-crosses—these last even in the midst of the market-places. All these are in ruins, it is true, but therefore the dearer to a tender and poetic people like these Irish Celts. When we think of all this, how silly, how absurd do these proselytizers appear! Why, if I were an Irish Catholic, I would treat these imbecile fanatics with contempt and scorn—by my sacred honor, I would!"

"And so they do, captain, so they really do," said Dixon. "That is precisely the feeling wherewith they are regarded by the nation at large, as far as I can see!—and no wonder—they bring it on themselves."

"But really, Frederick," said his wife, laughing heartily, "one would suppose you were half a Catholic yourself. Where in the world did you

pick up so much knowledge about this Ireland?—
I'm sure I wouldn't bother my brains about it, for
it is not worth half the trouble that's taken with
it! If it depended on me, the Irish might have
their religion, and welcome!"

"Not a doubt of it, Caroline," replied the captain—"and I don't think you are far wrong. As
to your wonder at my knowing anything of Irish
history, we'll let that pass, for any one who knows
you would never dream of *your* burthening your
memory with anything relating to Ireland. I only
want to set you and this good company right about
my probable tendency to Catholicity. No! no!—
it is a religion that would never do for me, because
of its various mortifications and humiliations. I
respect it, I confess, but, by George! I'd rather
see any one else embrace its tenets than myself.
If I were some thirty years older, then, indeed, I
would have less objection, but *now*"—he shook his
head with comical gravity—then starting to his
feet, led Eleanor to the piano, saying—"Pshaw!
what a dull subject we have been harping on for
the last half hour!—Do, pray, Miss Ousely, let
us have some enlivening music. You play Belli-
ni's grand marches, do you not?" Eleanor smiled
assent, and the whole company was soon listening
entranced to the "witchery of sweet sounds."

By the time the march was concluded, Ousely
made his appearance, and announced that he had

at last got rid of O'Hagarty. "And a d——d bore he is, too. I wish the Protection Society would send us a better specimen of a converted priest—I begin to despise this fellow, curse him!"

"I rather think," said Dixon, archly, "that it isn't the Society's fault—if they had better, they'd send better, that's all. You must only take him as you find him, for if you wait for a good, moral, intelligent *priest* from the Protection Society, you'll wait a long time, I can tell you. Such priests are only to be found *in* the Church of Rome—they never *leave* it."

Ousely was about to make an angry retort, when Trelawney proposed a game at whist, in compliance with a significant gesture from Mrs. Ousely. Seeing, however, that Eleanor and Amelia were looking over a volume of engravings, he contrived to be left out, and joined the young ladies.

"I thought you were going to take a hand!" said Amelia, pushing a chair towards him. "It was a pretty thing for you to propose cards, and then take yourself off. I fancy we have the pole-star somewhere about here; eh, Eleanor! what do you think?"

"I really don't know," said Eleanor, though her conscious blush spoke a different language. "I have not been accustomed to consider the astronomical bearings of this room." Just then as

eyes met Trelawney's and the blush deepened on
her cheek. Amelia smiled and shook her head.

"Well! well, good people, I'll be generous for
once. What did you think of Captain Hampton's
defence of Popery, cousin Trelawney?"

"I thought it very creditable to his head and
heart," replied the baronet; "he has read more
and thought more than one would suppose. By
the bye! Miss Ousely—"

"Nonsense!" cried Amelia; "why don't you
call her Eleanor, as I do? You may as well break
the ice at once!—how very ceremonious you are
with your *Miss Ousely!*" And she imitated his
tone so perfectly, that the others laughed heartily.

"Well!" said Trelawney, "I was going to ask,
when you stopped me, whether there were any of
those old monasteries in this neighborhood. I
should like, of all things, to see some of them."

"You need not wish long, then," said Eleanor,
"for we have one at Loughrea, within a few hours'
ride of us. There is an old Carmelite monastery
there, which dates back to the first years of the
fourteenth century. It is a very interesting relic
of the past greatness of Ireland, and is well wor
thy of attention, as a specimen of the ecclesiastical
architecture of that period. We can make up a
party and go there, the first fine day that comes."

"You will oblige me by doing so," said Trelawney, "as I may not soon have an opportunity of

seeing such a sight, and it will give me real pleasure."

"It will be a mournful pleasure, I warn you,"
said Eleanor, " for I defy any one to spend an hour
there without falling into a meditative mood. Even
our Amelia here—wild girl that she is—"

"Thank you kindly!" said Amelia, with mock
gravity; "but I'm not very fond of meditating,
like Hervey, ' among the tombs'—I leave that to
you serious people. Still, if you think of visiting
Loughrea Abbey, I have no objection to be of the
party. What do you think of asking the Reverend
Mr. O'Hagarty?" she suddenly added, with a smile.

"I rather think," said Trelawney, "that the
excellent gentleman is not much of an antiquary.
I should suppose him more interested in the respective qualities of Port and Claret, than in the
different styles of architecture, or the progressive
history of Christian art. But I see your father is
on the move, Amelia."

"I declare, so he is! I must be off and get on
my muffling!" So saying, away she ran, leaving
Eleanor and Trelawney *tete-a-tete* for a moment.
The only words that passed between them was a
whispered inquiry from Trelawney, as to where
the O'Dalys had taken shelter, and Eleanor's brief
reply that Phil Maguire had made his home theirs.
By this time the guests were all in motion, and
carriage after carriage rolled from the door.

CHAPTER XIII.

When man has shut the door, unkind,
On Pity, earth's divinest guest,
The wanderer never fails to find
A sweet abode in woman's breast.
CAROLINE.

"Pretty doings are here, sir, (he angrily cries,
While by dint of dark eyebrows he strives to look wise)—
'Tis a scheme of the Romanists, so help me God !"
MOORE's *Intercepted Letters*.

IT was on the second day after the ejectment of the O'Dalys that Sir James Trelawney rode over to Phil Maguire's, and as he gave his horse to a boy who was loitering around outside, those within the house were taken by surprise when he raised the latch and walked in. Phil and Nanny both came forward to welcome him, and Bernard O'Daly stood up from his comfortable seat in the chimney-corner to make a low bow to " the English gentleman—God bless him." Kathleen and Bridget got up from their spinning-wheels, and each dropped a low curtsey, and it was on every side, " God save you, sir!"—" I'm proud an' happy to see your honor here!"—" Will you please to take a seat, sir ?" But there was one smiling face there that arrested the young man's gaze for a moment—it was the face of Eleanor Ousely, who had been sitting beside Bernard, but had stood up with the rest.

" You here, Miss Ousely ?" he said, with marked emphasis.

" And I may retort," replied Eleanor, with her meaning smile; " Who would have thought of seeing *you* here ?"

" Oh, then, indeed, sir," said Bernard, " it's nothing new to see Miss Eleanor comin' amongst us. The Lord's blessin' be about her, she has been comin' to see us now an' then, ever since she was the size of our Eveleen there." This introduced Eveleen, who came modestly forward, at her father's bidding, to shake hands with " the strange gentleman."

When Sir James had said something civil to each of the others, he turned again to Eleanor. " But, surely, Miss Ousely, your father is not aware of this visit ?"

" Certainly not, Sir James ! but my mother is, and her sanction is quite sufficient for me. I have already told you," she added in a low voice, " that I am much interested in this family, and their present circumstances are truly pitiable. I know not what they should do were it not for Phil Maguire and his excellent wife. There must be something done for them, for they cannot be left as they are, and it may be some months yet before they can get relief from America. How I envy those," she said almost inaudibly, " who have available funds of their own !—But," raising her voice, " did you hear, Sir James, of the last visit which Bernard received from the Scripture-readers ?"

" No—when was it ?"

Bernard gave an account of the interview in his own simple manner, and as he proceeded there came a flush of indignation over Trelawney's fine features, and his dark eye sparkled with unwonted fire.

" The vile miscreants !" he exclaimed, when the old man had told all. " They would make the bitter cup more bitter still—surely they could have had no hopes of succeeding *then*—had they not often tried you before ?"

" Not *very* often, your honor," returned Bernard. " It was only once before that they ventured into the house, an' that was the night of Honora's wake. Poor Honora !" he added, rubbing the back of his hand across his eyes ; " it's well she wasn't alive to see or hear them !"

" But you may be sure they *had* hopes, your honor," observed Phil Maguire ; " for they sometimes *do* get people to give in at sich times that never would listen to them before. It doesn't happen very often, to be sure, but then they know very well that it's a hard trial an' a sore temptation for a father or mother of a family to go out on the wide world with their starvin' little ones ;

an' once in a while some poor creature gives in to
them for a start, just hopin' to keep the shelther
over them till something 'id turn up.   Oh, sir! sir!
if you only knew the twists and thricks of them fellows, an' the plans they take to get the poor miserable cratures hooked in!—an' still they go on and
on, though they see as plain as can be that they're
makin' no headway, nor gettin' no footin' in the
country—God forbid that they did!—sure they
know as well as we do, that no one goes over to
them only when they're jist in a state of starvation,
an' that as soon as ever they get any manes of
livin' they come back again where their hearts were
always.   Besides, it's well known, sir, both to *them*
an' every one else, that *death* brings every one
back—every one, your honor, that has time to send
for a priest.   Now isn't that a purty thing, sir! to
have these schamin' villains goin' about gettin'
money every where to convart the Papists, an'
makin' people b'lieve that they're doin' the world
and all.   I ask your pardon, sir, if I'm givin'
offence."

"Not at all, Mr. Maguire," replied Trelawney;
"you do but echo my own thoughts.   If *you* only
knew the sources whence this money is raised,
your surpriss would be still greater.   I believe
there is more sin committed in one day amongst
those who subscribe for the conversior of the Irish,
than there is in a whole year amongst your simple

hearted people. Shame on the hypocrites, and all honor to the virtuous poor, who brave every ill rather than give up their faith! But where is your son, Mr. O'Daly?"

"He's away at his work, your honor," returned Bernard; "when he can get it to do he's well pleased, poor fellow!" The old man sighed deeply, and there was a moment's silence, during which Eleanor arose, and taking Kathleen aside, put a small parcel into her hand, charging her to say nothing about it until she and Sir James were gone. She then went back to Bernard, and inquired what he proposed to do; "for," said she, "my mother is anxious to know."

"May the Lord bless her and you both, Miss Eleanor; and reward you for all your goodness to me an' mine! In regard to what I mean to do," he lowered his voice, "you know I can't stay here very long, so as soon as I get the childhren settled in some way I'll thry an' get into the poor-house!" The last word came out with a kind of sob, that told what words did not, the fearful anguish of the old man's heart.

"What's that you're sayin', Bernard?" cried Phil, whose quick ear caught the last word. "Now, if it's about the poor-house you're talkin', jist hould your tongue, for I tell you, honest man, tha' you an' I'll not be friends if you keep such a notion in your head."

"Well, but, Phil dear!" said Bernard, in a de
precating tone—"sure you know yourself that I
can't nor won't stay to be a burthen on you, an
me not able now to do e'er a turn at all.   For the
little time I have to be in it, it's no great matther
where I am."

"Now, Nanny, jist listen to that!" said Phil,
testily.   "Why, I think the man's takin' lave of
his senses.   An' indeed it wouldn't be much wonder if he did!" he added, in a sort of soliloquizing
tone.

"Tut, tut, Bernard!" exclaimed Nanny, stopping
her wheel for a moment.   "Now, sure, you know
well enough that you're welcome to stay here as
long as you live.   There's   room   enough for
us all!"

"The short an' the long of it is!" cried Phil
"that I wish I might catch you leavin' this to go
to the poor-house, that's all!   Upon my credit,
Bernard O'Daiy! it 'id go to the strongest man
between us—bad cess to me, but it would—an'
then I'd be sure to have it, so you may just as well
content yourself where you are.   You shan't leave
this house until you have one of your own to go
to, let that be when it may!   Humph! I declare
it's purty work I have with you!"

Eleanor and Trelawney exchanged glances, and
the latter, taking hold of Phil's rough hand, shook
it warmly   "You make me proud of human na

ture, Mr. Maguire!" said he struggling to keep in
the tear which moistened his eye-lid.

"Anan?" said Phil, who scarcely understood
his meaning, but probably guessing that it was
complimentary to himself, he went over to Eleanor, and began to give her an account of granny
Mulligan's achievement on the memorable day of
the ejectment.

"I heard of it before," said Eleanor. "But I
forgot to ask for the good old woman. Where is
she now?"

"She's gone down to Tullyallen the day," replied Phil, "jist to see how her daughter's grave
looks—wherever she is, she always goes there once
a month or so, to say some prayers over her *colleen
bawn*, as she calls her, an' to see the good man that
helped her to bury her."

"Well!" said Eleanor, "I must go now—I have
staid longer than I intended." She reached her
hand to Trelawney; "Good bye, Sir James! I
hope you are coming to see us soon."

"Will you not permit me to see you home,
now?" was the reply. "I wish you would."

"No, no, I must take what we call a near-cut,"
she replied with a smile; "I must scamper through
the fields, lest I might chance to encounter my father, who, of course, does not know of this visit.
I thank you all the same as though I could avail
myself of your offer."

She then shook Bernard by the hand, and as she bent to whisper some words of comfort, Trelawney murmured to himself, in the language of Shakspeare :

> "Kindness in women, not their beauteous looks,
> Shall win my love.'

Whilst he stood looking after her retiring form, Eleanor turned back from the door, to ask him whether he returned immediately to Clareview.

"No," said he ; "as you will not permit me to see you home, I shall call on Father O'Driscoll— a visit to him is *one* of my greatest pleasures. However, if you have any message to send, I shall be but too happy to take it." The message was for Amelia, and having given it, Eleanor hurried away, eager to escape hearing the prayers and blessings so profusely poured forth for her. What most struck Sir James was Eveleen's fervent exclamation : "Father dear ! isn't it a pity Miss Eleanor's not a Catholic ?"

"Husht, child, husht !" said Bernard, with a glance at Trelawney. "We must wait for God's good time—He knows best *when* to do an' *what* to do !"

These words made an impression on Trelawney that he did not soon forget, and as he rode along to Father O'Driscoll's cottage, they recurred often to his mind, and awoke a train of serious thought. He had gone about half the way, when he was overtaken by Mr. Ousely and the Reverend Mr O'Hagarty on horseback; they came up at full speed, but slackened their pace to have a chat with the baronet.

"We are just coming from the poor-house, Sir James!" said Ousely. "You must know that they have made me chairman of the Board of Guardians, and a d——d troublesome office it is, too—so this is our day of meeting, and I had to attend, whether I would or not. My friend O'Hagarty went with me for company, though he made himself useful, too—eh, O'Hagarty!"

"Why, I did what I could," returned the quondam priest, "but not as much as I wished."

"Well, well! never mind—'the worse luck now the better again,' you know. You see, Sir James, we have the world and all of trouble with these confounded Papists. There's not a day we meet but we have some fuss or another about religion—some refractory member that can't be broken in. So to-day we got Mr. O'Hagarty to try his powers of persuasion on them, but, upon my honor! he got the worst of the battle, ha! ha! ha! It's bad enough, and still I can't help laughing at it. Why, they wouldn't hear a word from *him*, at all!"

"More fools they!" said O'Hagarty, with a sly leer at Trelawney. "They don't know what's good for them."

There was something in his tone that Ousely did

not like, and he said with a sneer and a hoarse laugh: "I find your reverence is not more successful in making converts than in making love!"

"What do you mean, Mr. Ousely?" said O'Hagarty, bristling up, his face almost purple. "I don't understand you!"

"Pooh! pooh! man, don't be in a passion, now! you understand me well enough!—it'll never do for us to quarrel—you crack jokes yourself sometimes, so you must give and take, by Jove! I say, Sir James! are you coming our way?"

"No, Mr. Ousely; I am going to Father O'Driscoll's. I wish I may find him at home."

"The devil you are!" cried Ousely, almost fiercely, while O'Hagarty started as though an adder had crossed his path. "And pray what takes you there?"

"Certain business which concerns myself only," said Trelawney, drawing himself up with that stately air, which he well knew how to assume when necessary. "Many a happy and profitable hour I spend with him, for in him I find the devoted Christian, the accomplished gentleman, and the profound scholar."

"Deuce take him!" exclaimed Ousely, in a lower tone than was usual to him.

"Sir?" said Trelawney.

"I say, Sir James, that I don't understand this

thing of associating with Popish priests, except they do as my friend on the right has done!"

"Well, Mr. Ousely, our opinions on this subject are very different, and no good can come of our discussing it farther. I hope the ladies are well to-day!"

"Quite well, thank you. O'Hagarty! let us pull out—McGilligan is waiting for us before now! Good morning, Sir James! we won't detain you longer."

"My respects to Father O'Driscoll, sir!" said O'Hagarty, with mock politeness.

"I am not accustomed to offer insult to any one, sir," replied the baronet, haughtily, "and I certainly shall not deliver your message!"

"What a d——d proud young fellow that is!" said Ousely to his companion, when they had left the baronet some distance behind.

"He's worse than pr'ud," returned O'Hagarty; "he's impudent."

"Oh! as to the impudence, I can't agree with you," said Ousely, quickly; "he's too much the gentleman ever to be impudent. I think he only served you right that time, after all. Come, now, old fellow! don't be angry. Come home and dine with me, and after that, we'll ride over to the glebe, and see if Mr. Henderson has got that money for you yet. I don't know what the Society's about, that it isn't come to hand before now!"

O'Hagarty brightened up at the prospect of a good dinner and better wine in store for him, and by the time they reached the Hall, he was ready for anything that might offer. They dined an hour earlier than usual, and what was very *unusual*, left the table not more than " half seas over." Telling the ladies that they were going to see Mr. Henderson on business, and that they need not expect them for some hours, " because they'd have to take a tumbler or two with Henderson," the two worthies sallied forth, under favor of a rising moon.

They talked gaily and loudly all the way down the avenue, and along the road for a considerable distance, till they were almost close to the Catholic Chapel, with its burying ground lying calm and still in the moon's soft light, almost every little mound shaded, and as it were protected, by its white cross. There was a moment's silence, during which the two horsemen drew closer together; then Ousely spoke, but his voice was husky: " What in the world do these Papists put the cross at their graves for ? To scare away the devils, I suppose—ha ! ha ! and *ditto* the one on the top of the spire—ahem ! it isn't such a bad notion after all ! But why the deuce don't you speak, O'Hagarty ? *Your* thoughts are all of money—all right, old fellow, ' *money makes the mare go,*' as the old proverb says !"

They had now passed the Chapel, and O'Ha

garty suddenly recovered his loquacity. "Why,
a plague on your reverence," said Ousely, "is it
afraid of the ghosts you were, or what came over
you just now ?"

"Mr. Ousely !" replied O'Hagarty, in a tone of
indignation, "I hope you don't suspect me of such
foll; as that ? Bad as I am, I'm not much troubled with fear. There are many other causes that
might keep a man silent at such a time."

"Well! I'm glad you're not afraid," said
Ousely, putting spurs to his horse, "for here's
another grave-yard right before us now. Let us
go on—the night is passing !" But O'Hagarty
was again silent, and his eye involuntarily wandered over the small cemetery. All there was
calm and silent as in the one just passed; indeed,
it was a prettier sight to look on, for there were
stately monuments, and white tombstones, and neat
headstones, but the cross was wanting : that sacred
emblem—that sign of hope to man—was no where
to be seen. Half drunk as he was, O'Hagarty
shuddered, and a cold chill crept over him. Once,
twice, did Ousely speak to him without obtaining
an answer, and at last he laid hold of his arm, and
shook it roughly. O'Hagarty started, and was
very near screaming aloud, but finding that it was
Ousely's hand that had grasped his arm. he affected
to laugh at his own absence of mind. and made a
desperate effort to appear gay.

Very soon the pair came in sight of Bernard O'Daly's desolate homestead, and then it was Ousely's turn to fall into a reverie, but his did not last long, and he was just giving his companion an animated and somewhat *burlesque* account of the scene which had recently occurred there, when the stillness of the night was rudely broken by the report of a pistol, a ball whizzed over the neck of Ousely's horse, and struck himself in the right arm.   Ousely's cry of anguish, O'Hagarty's scream of surprise and terror, and a wild shout of " Vengeance! vengeance!" from behind the hedge, went up together on the still night-air; and then a solitary figure was seen darting across the field. Ousely, wounded as he was, would have pursued the assassin, but from this he was dissuaded by O'Hagarty, who represented to him that there was but little chance of their overtaking the fugitive, who could easily sneak into some hole or corner, while he was incurring the greatest danger from loss of blood.   " The best thing we can do," said he, " is to return to the Hall—that is, if you feel sufficiently strong. If not, we had better go on to the glebe."

" Home! home, then," said Ousely; " I think I have strength enough for *that* journey—ah! that d——d O'Daly!—I knew the villain was in him to the backbone!—But—oh!—don't go so fast O'Hagarty!—But he'll swing for this—he shall, by all that's good, if every cursed Papist in the country was at his back!—Easy—easy—I can't keep up with you!"

O'Hagarty had tied his pocket-handkerchief on the wounded arm, but still the effusion of blood was going on, and by the time they reached the gate, Ousely was so exhausted that he could barely call out for Larry Colgan. The tall gate-keeper was not slow in making his appearance, and seeing his master back again so soon, with O'Hagarty supporting him on his horse, he cried out: "Why, Lord save us, what's the matther with your honor?"

"Open the gate, you devil's limb!" replied his master; "what do you stand gaping *there* for?— don't you see I'm wounded—by Jove, O'Hagarty! I'm afraid I'm done for!—The d——d villain!"

"Be composed, I beg of you!" said O'Hagarty; "it's not so bad as you imagine!"

"Oh, murdher! murdher!" cried Larry; "is it bleedin' your honor is?—oh, then! oh, then!— what came over you at all, or who did it?" Then, without waiting for an answer, he ran to the door, screaming at the top of his voice for Peggy: "Come out here, Peggy!—sure the masther's shot!—he's kilt, Peggy!"

"Hold your d——d tongue," said Ousely—"it's like yourself, one half too long."

By this time Peggy was out, wringing her hands, and crying:

"*Musha!* who done it, at all, at all?"

"It will soon be known and heard, who did it!" murmured Ousely, who was growing fainter every moment. "I think I'll stay in the gate-house, O'Hagarty, till there's a carriage sent down for me. Go up as fast as you can, and tell them to send the phaeton—it's the easiest."

When O'Hagarty reached the house, he did not ask to see the ladies until he had first given the necessary orders about the carriage, and while Ben was getting it ready, he went into one of the parlors, and sent up a message to the effect that he would be glad to see Mrs. or Miss Ousely for a moment. Eleanor was down in an instant, for, knowing that her father and O'Hagarty had gone out together, both she and her mother were alarmed by this message, and his returning alone. On hearing that her father had been wounded, and was unable to come home without assistance, she clasped her hands, and turned pale as death.

"Oh, my poor father!" she exclaimed. "This is just what I often feared!—The blow has fallen at last!—Tell me, Mr O'Hagarty, do you think his wound is likely to be dangerous?"

"I should hope not, Miss Ousely! it is only in the fleshy part of the arm, and such wounds are seldom. dangerous. I don't think there's any serious cause for alarm."

"Thank God!" cried Eleanor fervently, and with upraised hands. "Thank God, if it were only on my dear mother's account. I hope you have ordered the carriage, Mr. O'Hagarty?"

"Yes, yes, I think it's ready by this time—there's no time to be lost."

"Well, then, will you be kind enough to go down in it, so as to support my poor father. I should go myself, were it not that I must break the news to my mother, and prepare her for what is coming! Merciful God!" she murmured, as O'Hagarty left the room, "how retributive is thy justice! But oh! do not—do not call my poor—poor father away *now*—leave him time to repent, oh my God! and to profit by this fearful warning! Now for my task, to acquaint my dear mother of what has happened!" Then wiping away the tears which were trickling down her cheeks, she hastily ascended to her mother's dressing-room, where they had both been sitting. Mrs. Ousely met her daughter at the door, and eagerly demanded what had happened.

"I know there is something wrong," said she; "I know it very well, so you need not try to conceal it from me." Then, when the light fell on Eleanor's pale and agitated features. "ah! I knew

It—there *is* something. Eleanor! my child! tell me—what has happened to your father?"

"Sit down, dear mother, and be patient—things are not so bad as you seem to suppose. My father is wounded, but it is only a flesh-wound in the arm. You may be sure it is not very bad, when he sat his horse for better than a mile after it happened. He will be here in a few minutes—the carriage is gone down to Larry Colgan's for him."

Mrs. Ousely sank almost fainting on a seat, for her trembling limbs would no longer support her. She gasped for breath, and for some moments could not articulate a word, but after a little, her tears burst forth, and she wept for a few minutes in silence, Eleanor making no attempt to console her, well knowing that it was better to let her emotion exhaust itself. When she saw her a little calmer, she reminded her that her father's wound was not considered dangerous, and that, after all, they had the greatest reason to be thankful, inasmuch as the same shot might have proved fatal.

"But, Eleanor dear!" said her mother, wiping away her tears, "who could have fired this shot? or did you hear how it happened?"

"I heard little or nothing more than what I have told you," replied Eleanor; "unfortunately, my father has made himself so many enemies in the neighborhood, that it is hard to say who has done it. Still"—she paused, and there came a deeper shade of thought over her beautiful features—" it might be—but no! I cannot, cannot believe it!—they who fear God as they do, will never have resort to such means!"

' Eleanor !" said her mother earnestly, " tell me, for God's sake, who it is that you suspect? do you mean—"

"Hush, hush, mother! here they are—there comes the carriage!—let us go down stairs! lean on me, my dear mother—you can scarcely stand! For mercy's sake, be composed, or your agitation will make my father think himself worse than he really is !"

Ousely's voice was now heard in the breakfast parlor, calling "Hetty! Nell! where are you all? Hang it, are they all asleep, that they take it so easy ?"

" Here we are, father dear!" said Eleanor, as she supported her mother's tottering frame across the room to where he sat, or rather reclined, in a large arm-chair. The sight of their pale, anxious faces was enough, and the wounded man held out his hand as they approached. " There, there, Hetty! don't take it so bad! don't cry now; not a tear, either of you—it's not as bad as it might be, no thanks to that d——d bloody-minded villain for that! Do you hear, O'Hagarty! send or go yourself down to the Police Barracks, and tell

Captain Ramsay to send up a sergeant's guard
here at once—you can take one of the men with
you, and come round by the glebe and bring Henderson with you—he's a magistrate, you know.
I'll not sleep this night, till that scoundrel, O'Daly,
is lodged where he won't get out of for a while.
Go at once, O'Hagarty ; and you, Eleanor, send
off another messenger for Dr. Coleman."

O'Hagarty hesitated a moment, and Eleanor, as
though she read his thoughts, exclaimed : " Father !
are you *sure* it was O'Daly who fired at you ?—
oh ! be not rash in such a case !—the O'Dalys—
father or son—are the very last persons I would
suspect of such a crime !"

" Nonsense, girl !" cried her father, raising himself to a sitting posture ; " I tell you it *was* that
young scamp, O'Daly !—who else would it be ?—
tell me that now—and it was just opposite to
O'Daly's house that the miscreant had concealed
himself !"

" Mr. O'Hagarty !" said Eleanor, turning to
him, " *you* were with my father when the deed was
committed—what do *you* say ?—could *you* identify
the person who ran across the field after the shot
was fired ?"

" I really was too much shocked," returned
O'Hagarty, " to take particular notice of the man,
but your father says it was this O'Daly, and the
moonlight enabled him to recognize him."

Eleanor turned away in disgust, murmuring to herself: "What a hard-hearted wretch—he knows that the young man's life is at stake, and yet he speaks with the coolest indifference." Aloud she said : " But, father, only think of the excellent character borne by these O'Dalys—there are others who might just as well be suspected, if the eject-ment be your only reason for accusing Owen O'Daly. A young man brought up as he was, is not very likely to commit such a crime with cool deliberation."

An angry exclamation from her father made Eleanor stop short, and O'Hagarty coolly said, as he buttoned up his coat:

" You forget, Miss Ousely, that the Popish reli-gion is essentially hollow and deceitful—sanctifying all crimes, provided they answer a certain pur-pose—it seems you know little of Popery, my good young lady?"

"More than you would suppose, Mr. O'Hagarty!" replied Eleanor, in a significant tone, as she left the room to send off for the doctor.

Mrs. Ousely remained with her husband, who would not be satisfied till O'Hagarty was fairly started, telling him that the bird might be flown if he made any further delay.

' It may be too late even now!" said he, his wrath blazing up again, at the bare idea. ".Ride now for life and death, if you wish to retain my friendship. Take Jerry with you—there he is!"

When the doctor arrived, and had examined the wound, he ordered Mr. Ousely to be undressed and put to bed, but said there was little or no danger, provided the patient were kept quiet, and made to observe a strict regimen.

"You must live low for a few days, my dear sir!" said he; "but you need not grumble at that, I think, considering that you have escaped so easily. Mind and avoid all excitement—I have dressed your arm now, and I assure you it is no more than a scratch—if you only do as I bid you, it will be as well as ever in eight or ten days! Good night, Mrs. Ousely! I was going to give you my parting charge, but I suppose it is Miss Ousely who will be head nurse. Now, Miss Eleanor, you are to see that your father drinks nothing stronger than barley water or weak tea. And as for his eating, let it be dry toast or water gruel!"

"Why, d——n it, doctor, do you mean to starve me?" cried Ousely.

"No, my dear sir, I mean to cure you—keep cool and quiet now till I see you again. I must now wish you good night, for I am in a great hurry."

O'Hagarty lost no time in sending the police, and the peaceful inmates of Phil Maguire's house

were just on their knees, saying the Rosary of the
Blessed Virgin, when the sergeant knocked at the
door.

" Who's there ?" said Phil.

" A friend—open the door !"

" Why, then, you're late abroad, whoever you
are ! an' your voice is strange to me !—what are
you wantin' at this time o'night ?"

" Let me in and I'll tell you !" was the reply.

" God direct me what to do !" said Phil in an
under tone to those within.

" Open the door !" said the stern voice without ;
" I command you in the Queen's name !"

" The Lord save us !" said one and another.
" It's the police—what brings them here ?"

" Why, open the door, Phil," said Owen, going
towards the door. " Sure none of us has any
reason to be afraid. I suppose they're searching
for some one that they think may have taken refuge
here !"

" Well, I'll open it, in the name of God," said
Phil. He did, and the sergeant walked in, followed
by a few of his men, the rest remaining outside.

" Fine night, sir !" said Phil. The sergeant
nodded in silence, and looking around, fixed his
eye on young O'Daly.

" Are you Eugene, otherwise Owen O'Daly ?"

" That's my name, sir !" replied Owen quickly

"I arrest you, then, in the Queen's name!" and he laid his hand on the young man's shoulder.

The women screamed aloud, and Bernard staggered forward, pale as death:

"What's that you say?" he stammered out.

"For *what* do you arrest me?" said Owen, with a firmness beyond his years. "What have I done?"

"Ay! what has he done?" cried Phil Maguire, as soon as he had recovered from the astounding effect of the sergeant's words. "I know he hasn't done anything to be arrested for—that's plain—but what *is* he arrested for?"

The sergeant looked from one to the other with his cold, dull eyes; then answered them all at the same time:

"He is arrested on suspicion of having fired at Mr. Ousely of Ousely Hall!"

"The Lord save us!" cried Phil—Bernard was not able to speak. "An' was Misther Ousely shot?—arrah, when did it happen, if you please, sir?"

"Come! come! I can't stand here answering questions. Put on your hat, young man! and come with us—you'll soon know all about it!"

"Sir!" said Owen, drawing his slight figure up to its fullest height; "Sir! I have never fired at any man, and if Mr. Ousely has been shot, I never heard of it till this moment. I have neither act, part, nor knowledge of it. When did it happen?"

"To-night—about an hour ago!" replied the ser
geant sternly. "Stephenson! have you the handcuffs there?—give them here!"

"Why, the Lord bless you, sir," cried Nanny
Maguire, "sure we can every one of us swear that
the poor boy didn't cross that threshold since night-
fall—we can, indeed, sir!"

"It's the thruth she's tellin' you, said Phil, ear
nestly; "we can take our Bible-oath of it. Why,
what in the world wide put it in any one's head to
accuse *him* of it—him that wouldn't hurt a dog!—
hut! tut!"

By this time poor Bernard began to realize the
dreadful truth!—they were putting the handcuffs
on his innocent child—his poor boy, that never
did man or mortal any harm!

"Oh, sir, dear!" he cried, the tears streaming
down his furrowed cheeks—"Oh, sir, dear, don't
do it—God for ever bless you, an' don't—oh,
Kathleen, Bridget, come here—an' little Eveleen!
all o' you come, childhren, an' beg o' the gentleman not to take your brother away from us. Oh!
sure he's all we have *now!*"

But neither tears nor prayers could avail—the
old man and the weeping girls, and Nanny, with
her officious kindness, were in turn pushed aside,
and poor Owen was marched away like a commo/
felon between two of the policemen.

CHAPTER XIV.

"Yes—rather plunge me back in Pagan night,
And take my chance with Socrates for bliss,
Than be a Christian of a faith like this,
Which builds on heavenly cant its earthly sway,
And in a convert mourns to lose a prey."—Moore

THE whole neighborhood for miles around was
thrown into consternation by the news of Ousely's
mishap, and O'Daly's arrest, consequent thereon.
The whole corps of the proselytizers was filled
with a holy horror, and sputtered out a great deal
of bile against the atrocious system, which not
only tolerated, but encouraged, such murderous
deeds. Some of them even talked of packing
up and decamping; for when such a man as Mr.
Harrington Ousely—a resident landlord, spending
his income liberally amongst his tenantry—when
he had been shot at, what could *they* expect?—
they, who were strangers in the country, and so
vilely misrepresented and misunderstood by the
ungrateful people for whose spiritual welfare they
were so exceedingly anxious. Truly, it was as
much as a man's life was worth to venture out
amongst such a set of savages. Others thought
that there was the greater field for their civilizing

exertions—the deeper and darker the shades of
Popery and its attendant vices, the more loudly
were *they*, the world's enlighteners, called upon to
remain, and to redouble their efforts to disseminate
Gospel truth, and to propagate sentiments of
Christian charity. As for this vile assasin, O'Daly
the holy conclave trusted he would be made an
example of, in order to deter others from attempt-
ing similar crimes. "It will be," said they, "a
crushing blow for Popery if he is hung, seeing that
these O'Dalys are considered as very pious, good
Papists. It is the best use the young ruffian can
be put to, for it may help to turn many away from
following ' the great delusion.' "

Such were the characteristic thoughts and say-
ings of the Scripture-readers and their employers,
but, by the country at large, the matter was
viewed in a far different light. Those who knew
the O'Daly family scouted the bare possibility of
Owen's having been guilty of such a crime, and
even went so far as to say that it was much more
likely that some of Ousely's *own kidney* had fired
the shot, for the diabolical purpose of having it
blamed on the Papists. Even those who knew
the O'Dalys only by repute, were deeply interested
in Owen's fate, and had but little sympathy for
the wounded man, who, of late years, was little
better than a public scourge, whether in his capa-
city of landlord or of magistrate. "The devil's

good cure to him!" was the brief but expressive
comment of by far the greater number. "It's
long since he earned that, and worse if he got it—
many's the poor family he sent to desolation, since
the unlucky day that he took it into his head to
join the Bible-readers!" "Yes. but poor O'Daly,"
said others; "I'm afeard it'll go hard with him,
whether he did it or not, for there'll be no want of
swearin'—the Lord deliver the poor *gossoon* out o'
their hands, if it's His holy will this day!"

"Amen! I pray God, in case he's innocent, an',
between you an' me, if he *did* do it, it's not much
to be wondhered at, considerin' what happened the
other day."

Such was the state of public feeling, on the day
that poor Owen O'Daly was sent off to Galway
jail, there to remain till the Spring Assizes. As a
special act of favor, his father and Kathleen had
been permitted to see him, but Father O'Driscoll
was refused admission, though the poor lad ear
nestly desired to see him. In vain did the priest
apply in person to the magistrates, the answer was
a cold, contemptuous refusal, and the prisoner was
sent off without the comfort of seeing his pastor,
or obtaining his parting blessing. This was "th'
unkindest cut of all" to poor Bernard; he and his
daughters, with Phil and Nanny Maguire, took
their station as near as they would be allowed to

the door, so as to exchange a sad farewell with Owen, who looked

> "As pale and wan
> As him who saw the spectre-hound in Man."

But he was calm and composed—*he* shed no .ears, though he could scarcely restrain them, when he saw his aged father and his three sisters weeping, but all unmanly softness was banished from his young heart, when he was rudely prevented from answering Phil's friendly greeting, and Nanny's fervent " God be with you, Owen *machree!*" Little Eveleen stretched out her arms to her brother as soon as he appeared, but she was pushed back by a policeman. " Owen, Owen dear !" cried the affectionate child, " sure you're not going away from us? sure you'll not leave us ?" A melan. choly smile was the only answer poor Owen could give her, and that smile only served to increase the anguish of the sorrow-stricken group.

" Well, I vow to God !" said Phil, dashing away the tear which he did not wish any one to see ; " I vow to God, this is enough to turn a man's blood into gall, but never mind, Bernard, never mind ! leave it all in the hands of God, an' you'll see that He'll bring Owen safe back to you. *He* knows who's innocent an' who's guilty, blessed be his name for ever. Come away *home*, Bernard—here, lean on my arm—keep up your head like a man— now, don't you know very well that all the Jumpers

and Bible-readers, and peelers in the country can't
hurt a hair of his head without it's God's will?"

"What's that you're saying about peelers?"
said one of the policemen, who was sitting on the
window sill.

"What's that to you?" replied Phil, bluntly;
" I'm mindin' my business, do you mind yours, that's
if you have any! Come, Bernard! Nanny, bring
the girls with you." The discomfited policeman
hurled an impotent curse after the sturdy farmer,
out, as Phil said, " he might as well whistle jigs to
a mile-stone, for all *he* cared."

The girls were profuse in their lamentations all
the way home, but the heart-broken father was
scarcely heard to speak. His sorrow was too
deep for words, and he could neither weep nor
complain. When they reached home, they found
Father O'Driscoll waiting for them, anxious to
offer some consolation to that afflicted family.

"So you have seen poor Owen?" said he.

"Och, *fareer gar!* yis, your reverence," replied
Bernard, "an' for me, I've seer the last of him, for
my coorse is nearly run, Father O'Driscoll, an' I'll
be at rest, I hope in God, before the 'Sizes comes."

"Ho, ho!" said the priest, in as cheerful a tone
as he could assume, "don't give up so easily, Bernard. Please God, you'll live to see Owen at
home again, safe and sound, and perhaps Cormac
and Daniel, too I just came now with some good

news to you. You must be quiet, however, before
I tell you a word of it."

"Oh, Father O'Driscoll dear, what is it?" cried
Bernard; "you see I'm as quiet as can be, now!"

The whole family gathered round in eager
expectation, and the priest smiled, as he glanced
from one anxious face to the other. "Now, what
I am about to tell you," said he, "must be kept a
secret amongst ourselves for some days longer.
I have heard something this morning, that, if true,
will extricate poor Owen from his dangerous position. There is a person who left yesterday in
great haste for Galway, there to take shipping for
America, and, from certain circumstances which
have come to my knowledge, it was he who fired
at Mr. Ousely."

"The Lord in heaven be praised!" cried Bernard,
clasping his hands in an ecstasy of gratitude.
"That news has made me twenty years younger,
and I think I could walk every foot of the road to
Galway, to tell it to my poor boy!"

"Yes, but you must remember what I told you,"
said Father O'Driscoll; "you're not to say a word
to any one about this, until I give you leave. It
might put our enemies on their guard, and it is
better to say nothing about it until we are quite
sure. I know myself that Owen is innocent—of
that I have no doubt whatever—but *my* knowing
it is of no avail, unless we have positive proof as

to who *is* guilty. I merely told you this in order to give you some reasonable grounds for hope."

"Well, God bless *you*, at any rate, Father O'Driscoll!" said Phil. "It's you that's always bringin' us comfort in one way or another. Won't you stay an' have some dinner with us, your reverence?"

"*Do*, Father O'Driscoll," said Nanny, who was bustling about in her culinary affairs, assisted by Bridget O'Daly; "there's a fine piece of bacon there in the pot, that's as sweet as a nut, an' some fine white cabbage that you didn't see the beat of this year."

"I wish I could avail myself of your kind invitation," said the priest, with a smile, "but, tempting as your bill of fare certainly is, Mrs. Maguire, I cannot wait for dinner. I have to go down to the lake shore, to see a poor woman who is lying sick there under a shed. She is a poor lone widow, who was turned out of her little place a fortnight ago, and since then she has been lying under a shed which the neighbors put up for her. Alas! such scenes are so common now-a-days," he added, in a sorrowful tone, "that they excite no surprise. But God sees the suffering of his people, and He will reward them! Well! Eveleen, my child! did you hem those handkerchiefs for me?"

"I did, sir," said Eveleen, coming modestly forward, with a small parcel in her hand.

"She was just waiting for you to ask, your reverence," said Kathleen; "she had them done two or three days ago."

"Indeed!" said the priest, laying his hand on the little girl's head. "Well! Eveleen! here's something to buy yourself a bonnet, or whatever you like, and I'm very glad to find that you are so industrious. I must speak to some gentlemen of my acquaintance, and get you some more handker chiefs to hem."

"Thank you, sir," said Eveleen, with a low ourtsey, and a bright smile of joy on her fair face. "But it isn't a bonnet or anything like that I'll buy with the money. I know myself what I'll do with it."

"And what is that, Eveleen?" asked Father O'Driscoll.

"A pair of shoes for my father, sir!" replied Eveleen, in a low voice, her face covered with blushes. "He's badly in want of them."

Her father would have stopped her, but it was too late, and the priest patted her head again, saying:

"You're a good girl, Eveleen, but don't be in a hurry buying the shoes. You haven't got enough there, and I'm sorry I haven't any more change. But there's a good time coming, Eveleen!" He then hurried away, leaving lighter hearts behind him than he himself had expected. So clastic is

the Irish—the Celtic heart! Before Eveleen had
got any one to see after the shoes, there came a
man to take her father's measure for a pair. At
first he would not tell who sent him, but when the
question was pushed home, he admitted that it was
Father O'Driscoll.

"Ar' God knows," added the honest shoemaker,
"he can ill afford buying for others, for, to my
knowledge, his own boots are none of the best—
I've mended them in one way or another five or
six times. But mind, you don't let on that I tould
you."

" May the Lord clothe his soul with the glory of
heaven," cried Nanny fervently.

"Amen, I pray God!" said Bernard; "and
yours too, Nanny!" for Nanny had knitted some
pairs of comfortable stockings for Bernard since
he had been her guest. "It's for which of you'll
do the most for us, anyhow," he added. "It's one
comfort we have, in all our throuble, that we've
plenty of good, kind friends—the Lord reward
them, here an' hereafther!"

Meanwhile, Mr. Ousely was rapidly recovering
from the effects of his wound. He was very soon
able to sit up, and to receive the congratulatory
visits of his friends, and his dressing-room was
crowded with visitors, for the first few mornings
after he was declared convalescent. It was in the
forenoon of that very day which saw Owen O'Daly

lodged iu Galway jail, that Sir James Trelawney
rode over to Ousely Hall. Befcre he went up
stairs, he had a short interview with Eleanor in
the breakfast parlor, and though he scarcely spoke
half a dozen words, they were sufficient to make
Eleanor's eyes sparkle, and her cheeks glow ; nay,
she even went so far as to reach out her hand
(which, it is needless to say, was warmly taken),
as she fervently exclaimed : " I give you joy !"

" But am I to be alone, Eleanor ?" said Trelawney, still holding the beautiful hand, and looking iu
the still more beautiful face.

" Not long, if God so pleases !" replied the
young lady, quickly. " But, go now—my father
will wonder why you stay, for your arrival has
been announced. You will find a bevy of spiritual
consolers with him—if you have any interior
wounds," she added archly, " you would do well to
lay them open for examination."

" The only interior wound I have," replied Trelawney, with his own peculiar smile, " is reserved
for other inspection than theirs. Next time I
come, I shall take the liberty of consulting *you* on
the subject." So saying, he turned away, leaving
Eleanor to construe his words as she best could.
When he entered the dressing-room, where Ousely
was seated in cushioned ease, he found himsel
face to face with O'Hagarty, and two other elderly
gentlemen, one remarkably tall, and the othe

remarkab.y short.   These were introduced respec
tively as the Rev. Mr. Henderson, and the Rev
Captain Wilson.   "Captain!" repeated Trelawney
to himself, "the *Reverend Captain!* what an odd
connection!"   Little knew he, poor simple youth,
of the strange anomalies of Irish life!   The
gentlemen, especially the two latter, "were de-
lighted," they said, "to make Sir James Trelawney's
acquaintance—they had frequently heard of him,
and had great pleasure in bidding him welcome to
Ireland!"   A formal bow was the only answer,
and Trelawney, having shook hands with Mr.
Ousely, and complimented him on his improved
appearance, took his seat on a couch near him, and
perceiving that his entrance had brought matters
to a dead stand, he begged that his appearance
might not interrupt the conversation.

   "Go on with what you were saying, Wilson!"
said Ousely; "Sir James, you know. is one of the
right sort."

   "I was just observing to our friends here," said
the reverend captain, "that force, physical force
alone, can ever make Protestants of these Irish.
We have been trying *every* other means for a number of years past, and the result is far from being
commensurate to the trouble and expense."

   "Physical force!" cried Ousely; "why, d——n
it, captain—I beg your pardon—what a discovery
you've made!—hasn't physical force been tried with

them for years and years before we began our
undertaking ?   By George ! if physical force would
convert them, they might have been converted
long ago."

"I quite agree with my friend Ousely," said the
tall rector; "I, for one, have more faith in the
effect of moral force; public opinion is the lever
that will upraise the heavy—the crushing weight
of Popery from off this unfortunate nation; bring
that to bear upon them, and our cause is sure to be
triumphant."

"Humph !" said Ousely, " all very fine talking,
but I should like to know how public opinion, or
moral force, call it which you will, is to be made
available in *this* case.   You might as well think to
apply it to the Hottentots, who, I take it, are just
as civilized and enlightened as the peasantry of
whom O'Connell, rat him ! was *so* proud.   Ha !
ha ! ha ! I wish he could only see them now !
But what do you say, O'Hagarty !—*you* should be
better able to form an opinion on this subject than
any of us !"

"My opinion is," said O'Hagarty, in a very dog-
matical manner, " that you should stick to the
soup and stirabout; leave the abstract questions
of physical force and moral force to be discussed
hereafter; but at the present time, when famine is
making such havoc amongst the people, you will
find the *eatables* all-powerful.   Bread, and soup,

and stirabout, my good friends," he added, looking
around with a scarcely perceptible sneer, " are the
only real weapons whereby you can defeat Popery,
and the time is exceedingly favorable—the Bible
itself is not half so powerful, take my word for it."

" Take care, my worthy friend," said Henderson,
with solemn gravity; " blaspheme not the Omni-
potent word of God !"

" Upon my word and honor," exclaimed Ousely
quickly, " I think what he says is perfectly true.
The only converts we *have* made were made by the
soup and stirabout, together with the other little
' creature comforts' in our gift. I really think that
what we have to do is to redouble our efforts to
get money, so as to enlarge our ' sphere of useful
ness,' as the saying is."

" Talking of money," said Mr. Henderson, ad
dressing the baronet, " I have not seen your name,
Sir James, on our list of subscribers. Surely you
cannot be insensible to the vast importance of the
work in which we are engaged ?"

" I confess I *am*," replied Trelawney drily. " I
cannot see its importance."

" How, sir ?" cried the reverend captain fiercely;
" do you pretend to say or insinuate that the people
are just as well as they are ?"

" I do, sir !—I think they are not only as well,
but much better as they are. They and their
fathers for countless generations have held the same

faith. I believe it has conducted millions of them
to heaven, and I see not why they should now be
called upon to give it up, or change it for another
of which they know nothing!"

This frank avowal took the worthy allies by surprise; not expecting such a home-thrust from such
a quarter, they scarce knew what to say, and could
only put on a swaggering air. Ousely put his
arms a-kimbo, and began to look fierce; the stout
clerico-militaro grew very red all of a sudden, and
Henderson knit his dark heavy brows into a very
formidable frown. O'Hagarty seemed to enjoy
the fun mightily, for still there was " the laughing
devil in his sneer," which Trelawney well understood.

" Really, my good sir," said Henderson, who
was the doctor of divinity amongst the saints of
those diggings, " it's very strange to hear such sentiments from an English Protestant." (Trelawney
smiled.) " Is it because the Irish have been grovel-
ling for ages in the darkness of superstition that
they are to be allowed to remain so? Their faith
is idolatrous, sir, as you ought to know, if you
know anything."

" And yet it is the very faith brought to them
by St. Patrick, fourteen centuries ago."

" I deny it, sir," exclaimed Henderson warmly;
" I deny that the present system, called the Popish
religion, is the same that St. Patrick taught. The

population of this island is very nearly as degraded now, religiously speaking, as it was when Patrick made his appearance on these shores. If his mission was *then* necessary, ours is just as necessary now !"

This was spoken with an air so triumphant that it was evidently considered unanswerable, and Ousely, accordingly, slapped his knee vehemently with his open palm, crying :

"Upon my honor! that's a clincher—eh, Trelawney? answer that if you can."

The reverend captain rubbed his hands in great glee, as much as to say, " He can't—do his best !"

Trelawney waited very quietly till the hubbub had somewhat subsided, then he said, with the utmost composure :

" There is one trifling difference, my worthy sir, between *you*" (bowing round to the three reverends) "and St. Patrick : *he* was sent by Pope Celestine, but pray who sent you to evangelize the Irish nation? By what authority do *you* come here to propose a new creed to the people ?"

" By the authority of God, sir, and in His name, accredited by his holy word !"

A scornful smile settled on Trelawney's features as he answered :

"Very well said, indeed, sir !—your answer sounds well as a rhetorical flourish, but it is scarcely satisfactory. Who is to vouch for your being

sent by God?—you say you come by His authority
but your saying it does not prove that it is so
Each one of you is his own ambassador, not the
ambassador of God, for if you be His ambassador,
where are your credentials?"

"The Bible, sir," replied Henderson, proudly;
"the Bible—no good Protestant requires other
credentials."

Trelawney smiled again. "Why, sir, if that be
so, you Anglicans have no sort of advantage over
any of the sects who have sprung from you—if the
Bible be your only credentials, then the Presbyterian, the Baptist, the Independent, the Unitarian,
has just as good a right as you have to undertake
the *conversion* of the Irish people from Popery."

He laid such an ironical emphasis on the word
conversion that it nettled his hearers beyond endurance. Ousely clenched his fist as though he
meant to inflict corporeal punishment on the offender; O'Hagarty's brow grew black as night, and his
face almost purple with rage, while the fat captain
got upon his legs, primed and loaded for a stormy
harangue. Henderson drew himself up, ditto his
shirt collar, then concentrating all the bitterness of
which he was capable (and it was no small amount)
into his look and tone, said, fixing his scowling
gaze on his smiling opponent:

"It seems to me, sir, that you argue much more
like a Papist, than a Protestant. Will you have the

goodness to set us right on that head? Are you
or are you not, a Protestant?"

"I *was*, when I came to Ireland—it is true I
never was an Exeter Hall Protestant, but still I was
sincere in protesting against something which I
had been taught to regard as the Church of Rome.
That was certainly my religion, if *protesting* can
ever be called a religion, but—"

"You protest no longer?" interrupted Henderson with a sneer. "You have learned to look more
favorably on the Church of Rome."

"So favorably, indeed," replied Trelawney,
coldly, "that I entered her communion this morning."

"The d—l you did?" cried Ousely. "Now, if
I thought you were in earnest, by all that's good,
I'd order you out of my house instantly."

"I shall not put you to that trouble, Mr. Harrington Ousely!" said Sir James, haughtily, as he
arose from his seat; "I am not in the habit of
jesting on serious subjects, and I repeat it, that I
had the happiness of being received into the true
Church this morning, by Father O'Driscoll."

Ousely sank back in his chair with a kind of
groan between a grunt and a sigh; he did not
dare to give full vent to his passion, when its
object was a gentleman of rank and fortune.
O'Hagarty shifted uneasily on his seat, and winced
beneath the contemptuous meaning of Trelawney's

glance. Henderson raised his hands and eyes in an ecstacy of pious horror; not so his fleshy and military brother, who could not refrain from showing his teeth, though he dared not bite.

"Perhaps you would be kind enough to inform us, sir," said he, in an ironical tone, "what were the arguments which induced you to go over to Rome ?"

"It would be too tedious to enumerate what they were, reverend sir," replied Trelawney; "but I can easily tell you what they were not:—they were neither bread, soup, nor stirabout! Mr. O'Hagarty's conclusive argument was *not* tried in *my* case; whether they or some similar inducements operated with *him*, I cannot pretend to say. Good morning, Mr. Ousely! good morning, gentlemen," bowing all round, "I am sorry to part such pleasant company, but *necessity*, you know, *has no law !*"

He was just leaving the room, when he heard Henderson saying; "I pity the young man, I do indeed!" whereupon he turned on his heel, and, holding the door half open in his hand, said, with keen irony, "I thank you, reverend sir, but I fear your compassion is thrown away, on one who has just left the religion of Luther and Henry the Eighth for that of Ignatius Loyola and Francis Xavier—my only sorrow is, for having so long remained out of that Church, which is, and has been the nursery of saints!" He bowed again,

and withdrew to tell Eleanor the result of the
conference. He found her with her mother, and
had made up his mind to say nothing at all about
it, but he had scarcely been seated, when Mrs.
Ousely said, very stiffly : " So it seems you have
become a Catholic, Sir James ?—you have kept
the process of your transition very quiet."

"Of my conversion, madam," suggested the
baronet, laughing at the odd substitute employed
by Mrs. Ousely; " pardon me for the liberty I take
in correcting you ; such a change is essentially a
conversion."

"Oh! as to that," observed Mrs. Ousely, " I
have not the slightest intention of entering into an
argument; your reasons for the change are, of
course, satisfactory to yourself, but I must own
that I have now less love than I ever had for Papists or their religion. It is not their fault that I
am not a sorrowful widow this day! I shudder
when I think of their hypocrisy !"

"Hypocrisy, my dear Mrs. Ousely!" said Sir
James; " I really do not understand you !"

"Why, how in the world could any one have
suspected those O'Dalys of such diabolical malice !
after them, no one need ever talk to me of Papist
morality or piety—they were cons'dered very
pious people—very pious people indeed, and just
see how far they carried their revenge—their cow
ardly treacherous revenge ! No, I shall never

again place confidence in Romish people—forgive me, Sir James, but I cannot help speaking as I do!"

"But, my dearest mother," said Eleanor, "you seem to take it for granted that young O'Daly did fire at my father. You are more severe than the British law, which always supposes a man innocent till he is proved guilty. You go on the opposite principle. Now, I have already told you that I do not believe O'Daly guilty; on the contrary, I am almost as sure of his innocence as if it were judicially proved. Time will tell which of us is right, but, in the meantime, I think we are not at all justified in condemning the Catholic religion, because one who professes it is suspected of having committed a crime. I need not ask, Sir James, what your opinion is?" said Eleanor with a smile; "I think I can guess it."

Trelawney started and colored. He had been thinking of something else, and it was his visible abstraction that made Eleanor smile.

"I beg your pardon," said he; "I believe I was forgetting myself, but certainly not 'to stone.' I was just thinking how unfortunate it is for me that your mother, Miss Ousely, is so prejudiced just now against Catholics."

"How so, Sir James?" said Mrs. Ousely, opening her eyes wide.

"I was in hopes, madam. that I should have had

your consent and good offices in a matter which is of vital importance to my happiness."

Mrs. Ousely was, for a moment, at a loss to understand his meaning, but one glance at her daughter's blushing face made it plain as the sun at noon-day. She was evidently taken by surprise, and her first emotion was one of displeasure: she sat upright in her chair, and put on a *very* serious look, and bit her lip till it became almost bloodless: gradually, however, there came a change in the expression of her features—they grew less and less rigid, until, at length, they resumed their usual mildness, and she said, in rather a kind tone:

"I cannot pretend to misunderstand you, Sir James Trelawney! and though I knew not before that you did my daughter the honor of thinking of her in that way, yet I will now frankly admit that I should have had no sort of objection to see Eleanor become your wife, provided she were satisfied, (she added with a smile,) but *now*"—she stopped and shook her head.

"I hope you do not mean to say, my dear madam," said Trelawney anxiously, "that *now* there is no hope?"

"I did not *say* so, Sir James! but I much fear that I might have said it. Even if I were disposed to consent, I am almost sure that Mr. Ousely never *would*   You surely have not now to learn that he

abhors the Church of Rome and—I had almost
said--all who belong to it."

Eleanor had turned away and pretended to be
very much engrossed by something which she saw
through the window.

"Eleanor, my dear!" said her mother, "come
here!" She turned, and her face was so pale that
it startled her mother, who hastily arose and went
over to her. "What is the matter, my dearest
child?" she said tenderly. "What have I said to
affect you so?"'

"My dear mother!" she said in a tremulous
voice, "it was merely a sudden faintness that came
over me—I am *quite* well now." And her blushing
cheek confirmed the assertion. Trelawney approached, and took her hand, which she made no
effort to withdraw. Mrs. Ousely looked from one
to the other, and then she sighed deeply.

"Miss Ousely—Eleanor!" said Trelawney; "I
now ask you, in your mother's presence—may I
still hope?"

Before Eleanor answered, she glanced at her
mother, and notwithstanding the unusual gravity of
her features, she saw that there was a smile lurking
around her lips. She raised her eyes to Trelawney's face, and said, with a smile so radiant that of
itself it might have inspired hope:

"Hope on, hope ever!" Then disengaging her
hand, she said:

"My dear mother! I leave you and Sir James *tête-a-tête* now, for I must go and see how my father is doing."

"There's an old woman out here wanting to see you, Miss Ousely!" said John, putting in his head.

"Do you know who she is, John?"

"Why, then, to be sure I do, Miss—it's granny Mulligan—sorra one else!"

"Oh, indeed!" said Eleanor; then, turning to Sir James, she asked if he had ever seen granny Mulligan.

"Yes, I saw her, if you remember, beside the death-bed of Mrs. O'Daly, but from all that I have since heard of her, I should like to see something of her."

"May I have her introduced, mother?" said Eleanor; her mother smiled assent, whereupon granny Mulligan was ushered in, much to her own surprise; "for," as she used to say, when telling the story, "it was the first time ever myself was in a parlor—an' for the matter o' that, the last time toc."

CHAPTER XV.

The cabless tigress in her jungle ranging,
  Is dreadful to the shepherd and his flock ;
The ocean, when its yeasty war is raging,
  Is awful to the vessel near the rock ;
But violent things will sooner bear assuaging
Than the stern, deep and wordless ire
Of a strong human heart.   BYRON.

WHEN granny Mulligan entered the room, she
threw back the hood of her red cloak, and looked
around with as much ease and self-possession as
though she were in Phil Maguire's kitchen.

"Your sarvint, ladies !" said she, nodding almost
familiarly. "Your sarvint, sir !" to the baronet,
who stood looking at her with a pleased smile on
his handsome features. "Miss Eleanor, dear, I
wanted to spake to yourself in private, but they
tell me you ordhered me in here. I ax your pardon for makin' so free, but you see it isn't my
faut !"

" Ce.. .mly not, granny ! and you are very welcome to come in. Will you take a seat ?"

" Oh no, Miss, thanks to you, I couldn't think of
sittin' down in this room—maybe the misthress
isn't plased with me for comin' in here ?'

Mrs. Ousely smiled, and said in her quiet way.

"Don't mind me, my good woman!—say what you want to say to my daughter!"

"Well, granny, and what is your business with me?" asked Eleanor, in her kindest tones.

"Well, I'll just tell you *that*, Miss. I came up here a purpose to ask you if *you* b'lieve this black lie against poor Owen? I'm tould the poor inno-cent *bouchal*'s in jail for firin' at your father—may they never—but I mustn't pray prayers on them, bad as they are! Now I wasn't about the place when they came to take the poor boy, or may I never do an ill turn, but I'd have given them a mark that they'd carry for a while—but when I came back that's the news they had for me, *inagh!* that Owen was lyin' in Galway jail!—Och! then *musha, musha!* what'll this world come to at all at all, when the likes of Owen O'Daly is taken an' clapped into jail for no raison at all. Miss Elea-nor! I ask you again do *you* b'lieve that he's guilty?" She strode up close to Eleanor, and looked up in her face as though she would there read the answer.

"No, granny," said Eleanor gravely, "I do *not* believe him guilty?"

"Then what's the raison that you didn't spake up for him?" exclaimed the excited old woman. "Tell me that now!—*you* could have saved the 'amily this last blow, an' you didn't do it."

"Granny Mulligan, you wrong n e!" replied Eleanor solemnly. "I *did* do my utmost, but my father was positive that Owen fired the shot, and Mr. O'Hagarty, who was with him, did not contradict him—what could I do ?—God knows I did all I could !"

The old woman was about to answer, when the door was thrown open, and the servant announced the Reverend Mr. O'Hagarty, the Reverend Mr. Henderson, and the Reverend Captain Wilson. Sir James, as they entered, drew back into the recess of a window, but kept his eye on the beggarwoman, anxious to see how she would acquit herself. Eleanor made a sign to her to leave the room, which she was in the act of doing when the captain caught a glimpse of her face under the hood which was now again over her head.

" Eh ! how is this ?" he cried; " stop there, good woman !—are not you the old lady who refused to go into the poor-house ?"

" Anan ?" said granny, becoming deaf all of a sudden.

" I say, aren't you old granny Mulligan ?" repeated Wilson, in a louder voice.

" I am !" replied granny, facing him ; " but you needn't spake as if you were in a mill, captain !— I'm not so deaf as all that comes to !"

Eleanor and Trelawney exchanged a merry

glance behind backs, and even Mrs. Ousely smiled
at the old woman's coolness.

"Don't be impudent, woman!" said Henderson.
"Remember who it is that speaks to you."

"Oh, of coorse!" said granny, in an ironical
tone; "I'll not forget *that!*"

"And pray what was your reason for refusing?"
said Wilson.

"My raison!" said granny; "Oh, bedad, I had
more than one raison!"

"Come! come! no quibbling!" said Henderson; "answer the question put to you!"

"I was jist goin' to do it, if you hadn't stopped me.
In the first place, I'd rather have my liberty than
be shut up in a prison, especially as I never done
anything to desarve it. Another raison is, that
I'm too fond of my belly to put myself in the way
of bein' starved; an' last of all, I'm tould there's
the divil to pay about religion in the poor-houses;
so, bedad, captain dear! I thought I had best stay
out, more betoken that the people made me welcome to a share of what little they had—the Lord
reward them for it!"

"But do you not know, my good woman?" said
Henderson, in a magisterial tone, "that begging is
now against the law?"

"Agin what law, Misther Henderson?" said the
old woman, with an air of great simplicity.

"Why, against the law of the land, to be sure!"

"Oh! if it's only that, your honor, we'll get over it—I was afeard you might have got some new laws from above," pointing upwards with her finger; "I know well enough that it's against the law to be poor now-a-days, for if it wasn't, sure there wouldn't be jails all over the courthry, for starvin' the life out o' the poor."

"How dare you speak so to me, you wretched woman?" cried Henderson, waxing wroth; "you know full well that no one is imprisoned without having committed some crime!"

"To be sure I do, your honor, I know it well enough—sure there's not one put into the jails. I mane, without bein' guilty of *poverty*, an' most o' them of another crime, that's even worse than that— *Popery*. Popery an' poverty, your honor, Popery an' starvation—them's the crimes that fills the poorhouses."

Trelawney drew farther back into the deep embrasure of the window, lest the reverend gentlemen should see him laughing, while Eleanor affected to be very busy indeed, assorting some silk in her work-box.

"Well, your honors," said granny, as she gathered her red cloak around her, "I'm for goin' now, if it's plasin' to the compary—bedad, it's quare company for granny Mulligan!" she said in an undertone, as if to herself.

"Do you know this gentleman?" said the rev

342 NEW LIGHTS; OR.

erend captain, pointing with an air of triumph to
O'Hagarty. The latter gentleman winced beneath
the keen and searching glance of the old woman.

"Do I know him, is it? ay, indeed do I, jist as
well as I want to know him."

"He was once a Romish priest, and if you
would only listen to him for a little while, he
would convince you, obstinate as you are!"

"Oh! may the Lord in heaven forbid that I'd
listen to him!" cried granny, with the utmost fervor. "Sweet Lord Jesus, stand between me an'
him!" and she crossed herself devoutly.

"I'll tell you what, now, my old hare!" began
O'Hagarty, his face flaming with anger, "I'll—"

"Don't spake to me!" cried granny, hurrying
to the door; "don't, I'll listen to any one, sooner
than you—your breath's unlucky, so it is!" and
pulling the door open, she darted out into the
passage, nor stopped till she got outside on the
lawn. Mrs. Ousely and her daughter both laughed
heartily, and Sir James, stepping forth from his
hiding-place, saluted the three gentlemen with
forced gravity. The two sanctimonious ministers
could not refrain from smiling, but O'Hagarty
looked as black as midnight, and taking out his
snuff-box, gave it a furious tap on the lid, as though
there were some vague connexion in his mind between it and granny Mulligan.

"I think you caught a Tartar, just now, Mr

O'Hagarty!" observed Trelawney, with as much composure as he could command. "The old woman seems to have no sort of reverence for your priestly character."

"Confound her for an old hag, I'll *make* her have reverence, or at least fear, some of these days, if she comes within reach of my horsewhip."

"For shame, reverend brother!" said Henderson, in his deep nasal twang, "why do you speak so uncharitably—the whip is not a proper argument—"

"It's the only one for the like of her!" returned O'Hagarty; "I tell you there's no use talking to these people; fill their bellies when they want it, and lash them like hounds when they're refractory, that's the only way."

"Just my idea," remarked Wilson; "these Papists are to be treated as were the Amalekites and Moabites of old—they harden their hearts against the Gospel, and scoff at the ministers of the Lord; therefore, I say, they deserve no mercy —but I beg your pardon, sir!" he said, turning suddenly to Sir James, who was standing talking to Mrs. Ousely. "I forgot that *you* were present, else I should not have expressed my opinion so freely."

Trelawney affected not to notice the insulting tone in which this was spoken, but he said with a bland smile "Pray make no apology, my good sir! your words are very consoling to me, I assure you."

"Consoling!—how is that?" cried Wilson, 'n surprise, for he certainly meant them to produce a far different effect.

"Why, they are truly consoling, inasmuch as they serve to convince me more and more of the infinite difference between religion and hypocrisy—between charity and cant—between the religion I have embraced, and the broken cistern I have rejected. Ladies, I must wish you good morning—gentlemen! your humble servant!" To Eleanor he said in a low voice, as he passed her: "*Adieu, au revoir.*"

"What a supercilious puppy he is!" said O'Hagarty, coming to the support of his crest-fallen friend.

"Pardon me, Mr. O'Hagarty!" said Mrs. Ously; "*I* see nothing *puppyish* or supercilious about the young man. I think him, on the contrary, by much the most finished gentleman I know."

"You must excuse my reverend brother, madam," observed Henderson; "his hatred of Popery sometimes carries him a little too far."

"So I perceive," said Mrs. Ously drily, and then the conversation dropped. The gentlemen soon after took their leave, much to Eleanor's satisfaction, as she felt anxious to go to her father whose patience was likely to be exhausted by that time.

About a week after, when Ousely was quite re-covered, his daughter took him into the front parlor one morning, telling him that she had something mportant to communicate.

"Well, Eleanor! what's in the wind now?" said he, as he established his rotund person in a cushioned arm-chair.

"My dear father!" said Eleanor, sitting down on an ottoman at his knee, "I have been requested to let you know that there is strong presumptive evidence—nay, more than presumptive—in favor of Owen O'Daly."

"And what the devil have I to do with their evidence?" cried Ousely. "What business have they sending me word about it? I know very well that the Papists are good at getting up plots, but what have I to do with them? The fellow's in jail for an attempt at murder—do *you* or do *they* expect me to interfere and get him out? Tell me that now, Nell!"

He spoke ironically, but Eleanor was no way discouraged. "And even if you did, my dear father, it would be greatly to your credit. If what the people say be true, I don't see how you can get over it!"

"And pray what *do* the people say?'

"They say—now don't be angry, father! that it was not Owen O'Daly who fired at you."

"Indeed!" said her father ironically; "and do they say who it *was*, then?"

"Yes, father, there's a rumor afloat that it was one of those unfortunate Ballyregan men who were evicted some weeks ago. There was one of them whose wife died on the road-side, you may remember."

"Confound them! didn't that fellow—I know who you mean—take himself off somewhere—he never showed his face here since. An unlucky villain he was, too, for I lost the Clifden races by him, and after all, his things weren't worth ten shillings, the whole lock, stock, and barrel—he'd do it with a heart and a half, I know, if he was in the country."

"Well, father, he *was* in the country—it seems the unfortunate man was down somewhere near Loughrea with a brother of his, ever since his poor wife's death, but he was observed lurking around on the very evening that you were fired at, and has not since been seen or heard of."

"The hang-dog ruffian!" cried Ousely; "I suppose he took good care to make himself scarce when he done the job—that is, if he did do it—but as we haven't *him*, we'll keep O'Daly, by ——, to see whether *he'll* be bagged or not—whichever of them done it, they're both well inclined—by the

Lord Harry! Nell Ousely, it's as likely as not that they both had a hand in it—there was murder in that scoundrel O'Daly's face on the day of the ejectment!"

"But, my dear father!" said Eleanor mildly; "you said yourself, and so did Mr. O'Hagarty, that there was but one man seen on that occasion!"

"Ay, but there might have been others still behind the hedge—we said there was only one who ran away, but there *might* have been many more, you know, Nell! who kept their ground."

"Well! well!" said Eleanor, "I see there is nothing to be gained by talking, so, with your leave, my dear father, we'll dismiss the subject for the present!"

Early in the afternoon Mr. Dixon rode up to the Hall, and after congratulating Mr. Ousely on his restored health, he said with a smile:

"My visit is not altogether one of friendship just now—it is partly on business, Ousely."

"Hang it, let us hear your business first, then!" cried Ousely, "so as to get it out of the way. Go into it, man, at once!"

"It appears," said Mr. Dixon, "that this lad O'Daly is innocent, after all, and, upon my honor, Ousely, I'm glad of it, for I had a great respect for that family, as far as I knew them, and was shocked to hear of any of them turning out so."

"And did you take the trouble of coming so far to tell me so?" said Ousely, abruptly. "You might have saved yourself the trouble, for my daughter Nell was beforehand with you. I don't care a d——n what the people say—let the law take its course!"

"But I have something more than hearsay to communicate, if you will only listen," replied Dixon, calmly.

"You have, eh? and what may it be?"

Eleanor, who was present, laid down her work to listen, and Mr. Dixon, after putting on his spectacles, and taking a pinch of *Lundy Foot*, drew a letter from his pocket, and commenced reading as follows :

"LIVERPOOL, October 29th, 18—.

"Your honor, Mister Dixon, as I often found you a good friend, the Lord reward you, I make free now to trouble you, hoping that you will excuse the liberty I take. You know as well as I do, honored sir, that I was put out of my little place by Mister Ousely some time back, and that my poor wife, God be good to her! died under a shed that I put over her in her sickness, to keep off some of the rain—she died there, sir, and left me, with four small children, without one to do a hand's turn for them—without a bit or a sup, except the cold water, and without a penny in my pocket. There's

no use in troubling you with a long story, sir, so
I'll say nothing of the state of mind I was in, nor
of the hunger, and cold, and nakedness myself and
the children were in after poor Ally's death--that's
not the thing that's on my mind now, you. honor,
so I'll pass it over. To tell the truth, I was bitter
enough again Mister Ousely, when I seen poor
Ally lying dead there on the road-side, and maybe
it was well for him that he didn't come across me
then. But, with the blessing of God, I got the
better of the devil that time, and went to my duty,
and got quite reconciled to bear everything with patience. There's a brother of mine that lives down
towards Loughrea, and he sent me word that if I'd
go to him, he could give me a shelter for a little
while, till times got better, only that I must try
and get the children into the poorhouse. Your
honor was good enough to apply yourself and get
them taken in, and so I went off to my brother's,
thinking that it was all right. Well, Mister Dixon,
I wasn't long away, till I heard that the guardians
were wanting the children to go to the Prodestan'
Church belonging to the poorhouse. With that, I
came up all the way to see the children, and to
warn them against these that wanted to ruin their
souls, and, sure enough the poor things promised
me that they wouldn't, on any account, go to the
Prodestan' meetin'. So I went my ways home,
but wasn't long settled there, till I heard that they

were forcin' the children to go to their Church—I
came up again, sir, and got to the poorhouse on
the very day that there was a meetin' of the
Board. Of course, I couldn't get in to know what
was goin' on, so I waited on the road abroad till
the meetin' was over, an' from where I was, outside the wall, I could hear the scrames of my poor
Johnny and Biddy—them's the two eldest, sir, and
I knew that they were batin' them bekase they
wouldn't consent to go to Church. Mister Dixon,
I'll not bother you with tellin' you how my blood
boiled, or what black thoughts was passin' through
my mind, but at last the doors were thrown open,
and the guardians—och, but they're the quare
guardians of the poor!—came steppin' out, and
myself had to step aside out of their way, but I
heard them talkin' as they passed by where I was
hid, and I found that it was Mister Ousely that
was the head of them all, and that it was him that
was hardest upon the children in regard of religion
Last of all, I found out that it was him that ordered
the cratures to be *flogged*, bekase they wouldn't go
to Church. Oh, your honor, when I heard *that*,
and remembered how it was him that in a manner
murdered my poor Ally, and drove the little ones
into the poorhouse, and then wanted to whip their
religion out of them, and rob them of the chance
of goin' to heaven, I declare to you, honored sir,
my brain got jist as if it was all on fire, and I

sure that before many hours passed I'd be revenged. The Lord forgive me! I know I was wrong; but then I couldn't help it, for it seemed as if I *must* do something, and I *did*, your honor! I borrowed a pistol from a friend of my own, and I watched for Ousely all day, till at last I got a chance, and I fired at him. Thank God, he wasn't killed, though then I intended nothin' else; but my hand trembled, and the ball didn't strike where I thought it would, and so he escaped *that* time. *Now* I'm glad of it; when the passion, or the madness, or whatever it was, cooled down, I thanked my God that I hadn't taken his life, though I know my sin was all the same. But what I trouble you for now, Mr. Dixon, is bekase I was tould that poor Owen O'Daly was taken up for what I had done, and as soon as I got my brother to lend me what would take me to America, I thought I'd write to you, sir, hopin' that you, bein' a magistrate, would set matters right. Of course I couldn't do it till I got here, but now the vessel's to sail in two hours, and then I'll be out of reach. Now, honored sir, I have tould you the whole truth, just as if I was goin' to face my God—I acknowledge myself guilty of this crime, and I ask God's pardon for it, but I can't rest so long as that poor innocent boy is in danger, or suspected of doin' what he never done, or never knew anything about. God bless you, Mister Dixon, and exert yourself for him and another thing I want to ask you, sir, for I know you're always willin' to do a good turn. Maybe you'd have the goodness to interfere for my poor children, and see that they don't persecute them for their religion. That's all the poor things have, and they may surely let them keep it As soon as I'm able, I'll send for them, with God's help. Don't forget to do what you can for poor O'Daly, your honor, and if a poor sinner like me can be heard, I'll never forget you in my prayers.   I remain, honored sir,

"Your most obedient and grateful servant,

"DARBY WHELAN."

During the time that Mr. Dixon was reading this epistle, Mr. Ousely made various exclamations, indicative of the various emotions to which it gave rise. Eleanor had kept an anxious eye on his movements, and saw that, just as she had expected, he was exasperated beyond measure. He could scarcely wait to hear the end till he struck his clenched fist on the arm of the chair, and exclaimed with the utmost vehemence:

"Darby Whelan may go be d——d, then, and every one that takes his part. D'ye hear that now, Dixon?"

"I do," replied Dixon, coolly and somewhat drily, "but I have no mind to put on the cap, for it certainly does not fit me. I have no part with this unfortunate Darby Whelan (though I must say that I never knew anything bad of him till this happened), but as *your* friend, Ousely, and the friend of justice, I come to have a talk with you on this subject. Miss Ousely, what do *you* think of all this? I know," he added with a kindly smile, " I know that you are never indifferent to the woes and sufferings of your fellow-creatures!"

" What's the use of asking her then?" said Ousely snappishly; "you know very well that Nell is always one of the 'friends of humanity, as the old song says. Come to the point at once, and let us hear what you expect from me."

Eleanor smiled, and nodded for Mr. Dixon to go on. " Well! Ousely," said he, " I'll tell you candidly what I think you should do. You should endeavor to get this young man liberated as soon as possible"—

" I'd see you and he—at Jericho first!" cried Ousely, interrupting him; " let the law take its course—that's all I say!"

" In that case," said Mr. Dixon, standing up, "it would be useless for me to insist farther. I am sorry for this, Mr. Ousely, even on your own account, for *I* cannot forget ' auld acquaintance,' though there are many now-a-days who take pleasure in doing so. Eleanor, my dear girl! good bye!" He held out his hand, but instead of taking it, Eleanor turned to her father, and begged him to think over the matter, before he gave Mr. Dixon such a flat refusal.

"I tell you I won't—mind your own business, Eleanor!—I don't thank any one for meddling with *mine*. You can all make a great rout about such fellows as Owen O'Daly and Darby Whelan, oh, certainly! they're not to be sneezed at, but Harrington Ousely may be shot like a dog from behind a fence, and even his own daughter has no sympathy for him—no resentment for the cowardly ruffian that did it. But, by ——," he swore an oath that made Eleanor shudder, "I'll send one of the rascals over the herring-brook, at least—curses on the villain! hanging would be too good for either of them!—the bloody Papist cut-throats!"

It was by a great effort that Eleanor maintained her composure while shaking hands with Mr. Dixon, who seeing her distress and confusion, said in a kind tone:

"Good bye, my dear! good bye. I hope your father will soon come to his senses—if he goes on in this way, the end of it will be the mad-house, take my word for it—that is, if some other persecuted creature do not take surer aim than Darby Whelan, which may God forbid. Give my best respects to your mother, Eleanor—it will be many a day before you see me in Ousely Hall again!—never except that man apologizes for his conduct, and comes himself to ask me!" So saying, he left the room and the house.

"Go you after him!" said Ousely, taking his daughter by the shoulder and thrusting her outside the door, which he slammed after her with a force that made the floor quiver. Eleanor's heart was like to break: it was the first time that ever she had received such treatment from her father, and she could scarcely persuade herself that the whole scene was not a dream. But alas! it was stern reality, and with all her filial affection, she could not help being ashamed of her father. Unwilling to tell her mother of what had happened, lest it might inflict a new wound on her already lacerated heart, she shut herself up in the privacy of her own apartment until she had obtained sufficient composure to meet her mother without any outward signs of agitation. Happy was it for her in that hour of trial that the light of true religion had lready dawned upon her mind.

Mr. Dixon was pursuing his homeward way at a pretty brisk pace, when, just as he reached the cross-roads where he had to turn off the highway, he discovered that his horse had lost a shoe, and was already somewhat lame. Fortunately, there was a forge about a hundred yards farther on, so he alighted, and led his horse by the bridle. On reaching the forge door, he came to a full stand, and could hardly refrain from laughing. The

blacksmith was hard at work shoeing a horse,
whose owner, a stout, sturdy farmer, was sitting
on a bench, waiting for the completion of the job.
By his side was Andrew McGilligan, his thin,
sharp features clearly defined in the bright glare
from the fire. He was holding forth, with his
usual circumlocutory eloquence, to the evident
amusement of his hearers, especially some three
or four ragged urchins, who were standing in a
group near the hearth, comforting the outward man
with the warmth of the blacksmith's fire, while
Andrew did his best to do as much for the inner
man, regaling them with plenteous draughts of
Scripture, and goodly quotations from various
tracts, concerning the abominations of Popery, *i. e.*
the scarlet woman, together with the manifold and
exceeding great blessings awaiting those who came
forth from " that unclean place," *i. e.* the Romish
Church.

"Thrue for you, Andy dear!" said the farmer,
with sly humor; "it's enough to make a body's
teeth wather, so it is, to look at the fine boilers of
soup an' stirabout that you have below there, an'
then at the school-house abroad, the beautiful
throughs that you're gettin' there to feed the children, athout any throuble at all to the creatures,
only to dip down their heads an' ate away : throth,
you may well say that there's blessins in store for
them that laves Popery!"

" More power there, Jack," cried one of the boys, and they all laughed heartily.

" Well done, Jack!" cried Vulcan, suspending his work for a moment, "well done, our side for a clane noggin."

Andrew scarcely knew whether to take it well or ill, but pretending to overlook the bitter irony of Jack's observation, he went on: " No, my very dear friends—I would I might say *brethren*—I do not allude to the things which concern the body, I speak of the things which appertain to the spirit! Oh !" said Andrew, in a fit of pious fervor ; " oh ! if you would only take the Bible in your own hands, as I have it now in mine, and read it even as I do, with a great desire to be enlightened, without asking leave of priest, or bishop, or pope, oh ! dearly beloved friends, but you would soon shake off the exceeding heavy yoke wherewith Rome has bound you."

"Ah, then, Andy dear," said the comical rogue, Jack, "would you plase to tell us what kind of a yoke that is, for the sorra bit o' *me* knows, though I was bred an' born in the Cath— ahem, I mane in the Church of Rome! Is it anything like the yoke that we put on the horses when they're ploughin', or maybe it's something like an ass's straddle, eh ?"

The roar of laughter which rang through the forge was re-echoed by Mr. Dixon outside, but his

laughter was drowned in the more obstreperous
mirth of the others, while his presence was concealed by the horse, which stood in the middle of
the forge.   As soon as the laughing had somewhat
subsided, Andrew spoke again, and as he spoke he
stood up.   "Ah!" said he, in dolorous accents, "if
you would only read this blessed book, you would
soon become meek and docile to the teaching of
those who would fain raise you from your degraded state.   This, my friends, this is the book which
overthrows the mighty power of Popery, and
tears away the veil that hides its deformity—this
is the sword wherewith we fight that monster, and
cut off his hideous head!   This—"

He was suddenly interrupted by the blacksmith,
who, putting down the horse's foot from off his
knee, snatched the volume from Andrew's hand,
and saying: "If that's the case, we'll give it warm
quarters!" he very coolly flung it into the fire.
Then, taking the horror-stricken Scripture-reader
by the back of the neck, he gave him a good shake
and put him out of the forge, telling him, as he
valued his bones, never to show his face there
again.   By this time, Mr. Dixon had got in by a
back door, and, as no one spoke to him about what
had happened, he affected not to have seen it, and,
telling the blacksmith that the groom should come
for the horse in the course of an hour, he set out
on foot for his own house.

CHAPTER XVI.

Think'st thou there is no tyranny but that
Of blood and chains ?   BYRON'S *Sardanapalus*

But happy they, the happiest of their kind,
Whom gentle stars unite, and in one fate
Their hearts, their fortunes, and their beings blend !
                              THOMSON'S *Seasons*.

It was fortunate for Owen O'Daly that Mr
Dixon was on the Grand Jury on the day when
his case was brought before that body. Ousely
as the complainant, was, of course, incapacitated
from "sitting," but he blustered and swore most
awfully, and did all he could to brow-beat his
brother jurors into a perfect conformity with his
own views. O'Hagarty was the only witness he
had to bring forward, and when that worthy gentleman came to be examined, though it was quite
evident that he wanted to oblige Mr. Ousely, yet,
do what he would, he could not *plump* it. Ousely,
'ndeed, had said that they were both ready to
swear that it was O'Daly whom they had seen
running across the field, that is, "to the best of
their knowledge," but O'Hagarty was too cunning
to "go the whole hog;" he knew very well that it

was *not* O'Daly, and what was more to the pur
pose, he knew that there was plenty of respectable
evidence to prove the lad's innocence, and he said
to himself, with his accustomed prudence, "It
will be an ugly thing if I am found out giving false
evidence—I know this Maguire is a touchy old
fellow, and has plenty of money to spend, and who
knows but it's an action for perjury they'd be
bringing against me. I'll tell you what it is, Bernard O'Hagarty! you'd best look out for number
one—everything depends on reputation, and I can't
afford to throw away the rag that's left me—let
Ousely swear as he likes, I'll not get myself into
trouble as long as I can help it." So out it came,
on the examination, that he could not swear positively.

"Oh, indeed!" said Mr. Dixon, who undertook
to cross-examine the reverend gentleman; "You
cannot swear positively that it was O'Daly—now,
I ask you, on your sacred oath, Mr. O'Hagarty,
are you not quite sure that it was *not* O'Daly?"
O'Hagarty looked down and was silent.

"Answer me, sir," said Dixon sternly, "on
your oath, was it not a much older man?"

"I—I—rather think so."

"Very good—that settles the matter. Gentlemen, I have done—does any of you wish to ex
amine the reverend gentleman?" Several of them
did, for the greater number were on Ousely's side·

sat the reverend Bernard had the prosecution for
perjury so constantly before his eyes that there
was nothing to be made of him, further than that
he had seen Mr. Ousely shot, and had seen a man
run from behind the hedge, but who the man was
he could not say.

The evidence for the defence was then brought
forward. Phil Maguire and his wife swore posi
tively that Owen O'Daly had not left their kitchen
from four o'clock on the evening in question until
he was taken away by the police. The father and
the two elder sisters of the prisoner were each
examined, and all agreed so perfectly in every particular that there was no getting over such a body
of evidence, especially when there was nothing conclusive on the opposite side. Last of all Mr.
Dixon read Darby Whelan's letter, and corrobo-
rated many of the statements therein made. This,
it would seem, was calculated to remove every
shadow of suspicion from O'Daly, and so Mr.
Dixon thought, but he reckoned without his host,
for he had no sooner finished the reading of the
letter than one of the jurors started to his feet, and
begged to suggest that the pistol which Whelan
had borrowed might have been furnished by
O'Daly, which, if done knowingly, made him an
accessory in the crime. The suggestion was eagerly
taken hold of (so true is it that the instincts of
Irish landlords are almost invariably *against* the

poor man and *for* the rich man), and Mr. Dixon
was under the necessity of summoning Phil Maguire and Kathleen O'Daly, to prove that Owen
had never owned a pistol, or indeed, fire-arms of
any kind.

"And now, gentlemen," said Mr. Dixon, "it
appears to me that we have not the shadow of an
excuse for bringing in a bill of indictment. I
know that you are all the personal friends of Mr.
Ousely;—so am I—but our friendship for him
ought not to interfere with the administration of
justice. If the real offender were before us, I
would be one of the first to agree to the finding of
the bills, but no rational man in our position can
shut his eyes to the fact that this poor lad is in no
way implicated in this crime,"

The result was that the bills were thrown out,
and an order was sent to the high sheriff to liberate
Owen O'Daly. When Ousely was informed of
the decision of the Grand Jury he was highly
offended, and swore that Popery was even getting
into the jury-room. But his anger was principally
directed against Dixon and O'Hagarty, the latter
of whom he pronounced "a d——d old humbug,
and a traitor to boot!"

It was Mr. Dixon himself who brought the news
to the anxious group without. They were sitting
on a bench in the hall, and when the worthy magistrate appeared on the stairs, they all stood up.

The old man trembled so that he could not stand without the support of Phil's arm, and it was that true-hearted friend that asked Mr. Dixon the question which was hovering on Bernard's lip.

" Well, Mister Dixon !" said Phil, " we're waitin on your honor, to see what sort of news you'd have for us."

" Good news! good news !" said Dixon, his honest face beaming with pleasure, as he reached his hand to the old man. "The bills are thrown out, Bernard, and the high sheriff has instructions to liberate Owen."

Phil and Nanny cried out, " The Lord in heaven be praised !"

"An' next to Him above, Misther Dixon, it's you we may thank for it !" said Phil.

Bernard could not speak for a moment, but he sank on his knees, still holding Mr. Dixon's hand, and the tears burst forth in torrents from his eyes. Mr. Dixon would have raised him, but he said : " No, sir—no, your honor ! I'll not rise till I thank you—thank you on my knees for what you've done for Owen, an' for us all. You've saved my gray hairs from what never came on them yet, with all our poverty, an' that's *disgrace*. May the Lord reward you this day an' forever more, an' may He grant you a happy death an' a favorable judgment. Amen. Now, Phil dear, help me up." Mr. Dixon coughed, and cleared his throat and then said : "Pooh, pooh, Bernard! I've done nothing but my duty !"

Kathleen and Bridget now came forward, and thanked Mr. Dixon for the share he had had in effecting Owen's liberation. "Why, upon my honor," said he, with a benevolent smile, "I cannot understand all this. Do let me away from here, for your gratitude is a heavy load to carry. Good bye! I shall see that Owen is speedily set at liberty."

When Ousely reached home after the examination, he bolted into the room where his wife and daughter were sitting at work, and threw himself almost breathless on a seat.

"Well! my dear, how did the investigation go?" said Mrs. Ousely, in her softest tones, while Eleanor glanced sideways at her father without speaking.

"Just as I might have expected," replied her husband ; "I might have known very well what was in that stupid load of flesh, O'Hagarty. He's the first converted priest ever I trusted anything to, and by the Lord Harry, he'll be the last. The confounded ass shirked out when it came to the point, and his evidence wasn't worth a brass far thing—he's not worth his room in a house—rat him ! I don't believe he's any more of a Protestant than I'm a Papist, only just to serve his purpose, the hypocritical knave !"

"Why, father," said Eleanor very demurely, "it is only a few weeks since I heard you call him 'the prince of good fellows'—'a merry old soul,' with ever so many other eulogistic epithets."

"Humph!" said her father, gruffly; "you heard me say more good of him than ever you'll hear me say again, that's one thing, I can tell you. The young vagabond has escaped for this time, thanks to O'Hagarty and Allan Dixon; but if I don't nab him yet, my name's not Harrington Ousely. He'll not get off like Darby Whelan, by h—— he shall not!"

Eleanor shuddered, but said nothing. She looked at her mother, and was shocked to see her leaning back in her chair, as pale as ashes. Seeing her daughter's consternation, Mrs. Ousely smiled, and made a sign to her to take no notice, that she would be better soon.

When the family assembled at dinner, Mrs. Ousely handed her husband a letter, which had been brought, about half an hour before, by one of Mr. Dixon's servants. Eleanor glanced at the letter, as she passed it down to her father, and her face was instantly covered with blushes. Ousely was opening the letter without a word, when his wife took the precaution of ordering the butler to withdraw, knowing that if it contained anything unpalatable, there was sure to be an explosion. Eleanor kept her eye on her father, and saw that

31*

... he read his color rose higher and higher, till his face was of a scarlet hue, and his eyes glowed like living coals.

"Eh ?—what ?" he cried, in a thick, husky voice, "it is only a week or two since he boasted to my face of having turned Papist, and now he has the impudence to ropose for my daughter! By the Lord Harry, Nell, I'd as soon you'd marry the devil—the graceless young scamp! did he dare to suppose for a moment that Harrington Ousely would let his only child go headlong into the gulf of Romanism—I say, Hetty, what do you think of that ?"

"Of what, my dear ?" said his wife, as though she had no idea of what he meant.

"Why, of that d——d English turn-coat proposing for our Nell ?"

"Well, since you *have* put the question to me, I suppose I must answer it. I think that Sir James Trelawney, with his high connections, ancient family, and a rent-roll of fourteen thousand a year, is a match for any woman in Ireland. As to his person and manners, and moral character, we know they are altogether unexceptionable. Of course, his going over to Rome is very unfortunate, but then—"

"But then!—zounds, Hetty! is *that* the way you take it!—I tell you if he were a prince of the blood, instead of what he is, he shouldn't marry a child of mine. If he had been always a Papist it wouldn't be half so bad—but a turncoat!—I say, Nell!" turning short round to his daughter, "it can't be possible that you gave him any encourage ment—did you, or did you not ?"

" I certainly did, sir !" replied Eleanor firmly, though her cheek was ashy pale. " I saw in him every quality which I could have desired in a hus band, and I had every reason to suppose that both you and my mother were favorably disposed to wards him."

" Favorably disposed !—to be sure we were be fore he thought fit to join the ranks of Popery !— a fellow that would let himself be drilled and tutored by a priest, and hoodwinked out of the use of his senses, is no son-in-law for me, and, by George! he's no husband for Eleanor Ousely !"

" But suppose, my dear father," said Eleanor calmly ; " suppose that his being a Catholic is not an insuperable obstacle with me—suppose I saw in his recent change of religion no reasonable grounds for revoking my consent, previously given—how would that be ?"

" How would that be ?" he repeated, mimicking her tone ; " why it would be that you might marry him, and go to the d——l if you chose, for me. You'd be no longer a daughter of mine—that's how it would be—if *you* haven't a spirit above

Popery ; *I* have, and you ouyht to know that—I despise turn-coats !"

" And yet my dear aunt Ormsby is a Catholic !" said Eleanor mildly ; " surely you do not despise her ?"

" I *do*, by H—— !" cried Ousely ; " I do despise her, and if I saw her I'd spit in her face—the low-lived jade ! there's not a drop of the Ousely blood in her, and I suspected as much long ago."

" My dear ! the dinner will be spoiled," said Mrs. Ousely ; " you had better drop this subject for the present."

" I won't drop it, Hetty !" exclaimed her husband, striking the table with his fist, " till Eleanor does one of two things ; she must either promise me to have nothing more to say to this Trelawney, or else acknowledge that she's a Papist at heart— there's no use in humbug—let her be either one thing or the other !"

Mrs. Ousely looked distressed, but Eleanor was perfectly calm and collected. She was not deceived by her father's coolness, unusual as it was with him : she knew that it proceeded from a fixed and settled purpose, and was merely assumed to draw out her real sentiments. But she quailed not before the storm ; she felt that the time for concealment had passed away, and that prevarication was no longer possible. Breathing an inward prayer for strength in this great trial, she said in a firm voice :

"I accept the alternative, father !—for I am sensible that the time for my confession of faith has come. I, too, am a Catholic—in heart and soul a Catholic !"

Mrs Ousely screamed, and clasped her hands with an instinctive fear for her daughter; the hot blood rushed to Ousely's face—his eyes flashed—his very lips trembled with passion, and his fingers worked convulsively; for some minutes, he sat glaring on his daughter in ominous silence—then his lips began to move, as though he were about to speak, but before he could get out a word, Eleanor came round the table, and knelt at his feet, saying :

" Father—*dear* father! I am truly grieved for having provoked your anger, but consider that the interest of my immortal soul was at stake, and then you *will* not—cannot blame me !—Forgive me, my dear, dear father, and I will never marry any one without your consent !"

Her mother, too, besought him not to be too hard on Eleanor. " You know, my dear !" said she, coming forward and laying her hand on his shoulder; " you know that this is a matter in which we have no right to interfere !"

" Don't talk to me about interfering, Hetty !" exclaimed her husband ; " why shouldn't I interfere ? Isn't it the greatest disgrace that ever came across me?—Here have I been these five years doing all I could to banish Popery from this neighborhood, and then to see my own daughter embracing its nonsensical tenets—d—n it! Hetty!—it's a burning shame, and I *won't* forgive her—no, by H——, I *never* will!—get up out of that, girl! and go to your room—don't leave it, either, till I order you!"

"I obey you, father!" said Eleanor, rising; "but remember tyranny may be carried so far that disobedience may become lawful!—justice and conscience are on my side—I leave you to consider what is on *yours!*" So saying, she walked quietly out of the room. Just as she had expected, her father was bewildered: he could not, by any means, understand the cool determination with which she spoke, and long after she had left the room, he sat staring at his wife, and she at him, in silent amazement.

"By the Lord Harry, Hetty!" said he at length, "that girl is a riddle that I can't make out. Why, she speaks with as much composure as though nothing were the matter. Do you really think she would go off without my leave to England?"

"I should not be at all surprised," replied his wife. desirous of making the most of his fears. "You know that Eleanor has a strength of mind that shrinks from no danger, and if she is once convinced that it is her *duty* to oppose your will,

she will do it at all hazards. She has been long
wishing to accept her aunt's invitation, and now
that she has made up her mind to become a Catholic, as her aunt has already done, I think she will
go to her, if you persist in your present course.
Then if she goes to England, she will, of course,
marry Trelawney with her aunt's sanction. Oh,
Harrington! think of it, I implore you; leave me
not a childless mother—if you drive Eleanor from
me, you will kill me outright!" The tears which
fell profusely from her eyes touched Ousely's
heart, hard as it sometimes was.

"Confound it, Hetty!" said he, quickly, "don't
you know it would be just as hard on myself to
part with Nell; but what can I do—tell me *that*,
now! How could I have the face to speak a word
against Popery, when every one knew that my
own daughter was a Papist,—hang it! if she'd
only keep it to herself, and not disgrace me before
the public!"

"But that she could not do, my dear," said his
wife mildly; "if she be a Catholic at all, she will
be one openly and aboveboard—you could expect
nothing else from Eleanor. But now tell me candidly, Harrington, are you afraid that she cannot
save her soul in the Church of Rome?"

"I'm afraid of no such thing!" he replied,
shortly; "of course she can save her soul in it—
why not? That's not the trouble, at all!"

Mrs. Ousely was silent. She was thinking of all the hard names she had so often heard her husband apply to the Church of Rome, and she could not help saying: "Bless my soul, then, if that be the case, what is the use of our missions? Why spend such vast sums of money in endeavoring to convert the Papists, if they can be saved as they are?"

"Hold your silly tongue, Hetty! you don't know what you're saying!" was the polite and most conclusive reply of Mr. Ousely, as he drew over to the table, and arranging his napkin, began to carve a magnificent goose which lay before him. Mrs. Ousely proposed sending up for Eleanor, but he laid his commands on her to do no such thing. alleging that there was nothing better for refractory people than solitary confinement. "Let her stay till she cools," said he, "perhaps she'll be a little more reasonable the next time I see her." His wife shook her head, but said no more on the subject.

When Eleanor reached her dressing-room, she sat down and wrote a note to Father O'Driscoll, of which the following is a copy.

" REVEREND AND DEAR SIR :

"I this day informed my father and mother of my having become a Catholic. My dear mother is not displeased with me—this I can see, though have since had no private conversation with her but my father is fully as much incensed against me as I had expected. He will not hear of my marrying Sir James Trelawney, because of his apostacy, as he chooses to consider it, though he admits that, in every other respect, he is just the man whom he would have chosen for me. Now, reverend sir, what I wish to ask you is this. Am I, or am I not, justified in giving my hand where my heart is long since given, and with the sanction of my dear, my excellent mother, in case my father is still obstinate in refusing, on the plea of religion? I shall leave the matter to your decision, as my spiritual guide and director."

In the course of the evening, she received the answer. Father O'Driscoll said that it was her duty to use every possible means, in order to obtain her father's consent; but that, in case he still held out, she was no longer bound to obey him, inasmuch as he forfeited the rights of a parent by endeavoring to coerce the conscience of his child. In that case, she must rest satisfied with her mother's approbation, and leave the rest to God, who, in his own good time, would move the heart of her father.

Eleanor communicated this to her mother, who was greatly distressed. She could not blame her daughter, nor yet Father O'Driscoll, but still she shrank from the prospect of losing that beloved child, who was indeed " all the world to her." She made a last effort to persuade her husband, but he was even more obstinate than before, and cut her short by declaring, with a tremendous oath, that he'd sooner see Eleanor Ousely in her grave, than see her marry a Papist—" though the hussy had the confounded impudence to tell me she was one herself. Let them go to blazes, and get married *if they like*, but they'll never be married *with my consent !*"

There was an emphasis laid on certain words in this sentence, which suggested a new train of ideas to Mrs. Ousely's mind, and though her husband looked as fierce as he could well do when he uttered these words, yet, in the course of ten minutes after, the good lady stole into her daughter's room and whispered : " Eleanor, my dear ! you may appoint an early day ; you have *my* consent, and my blessing, and that is all you want just now."

Eleanor looked inquiringly at her mother, but the latter put her finger on her lips, and merely said : " You must arrange it all with the Dixons— it is to be a private affair, you know—unknown to us !"

" But, mother," said Eleanor in a tremulous voice, " how *can* I leave you ?—what would you do without your Eleanor ?"

" Never mind *me*, Eleanor ! think only of your self at present. Hereafter we can easily manage to be together most of our time, either here or in your English home!—Go, my child—my beloved child—go, and God's blessing be with you!"

"Why, mother," said Eleanor with a faint smile, "that is just what a Catholic mother would say!"

"It is the natural outpouring of the mother's anxious love, Eleanor!"

Eleanor kissed her mother's forehead, and went in silence to answer a note which she had that morning received from Trelawney, and as she went, she said to herself:—"Yes—it is even so—the spirit of religion—the living, actuating spirit is essentially Catholic—whatever devotion, or genuine piety is still to be found amongst Protestants, can be clearly traced to a Catholic basis. Thanks be to God that he whom I have chosen for the partner of my future life has already sought and found the fulness of truth!"

On the second morning after this, Eleanor Ousely stole softly down stairs in the grey of the morning, and thought to have passed out unnoticed by any one. She had taken leave of her mother over-night, and was not aware that any other in the house suspected what was going on. To her great surprise she found the servants assembled in the hall, to wish her "good luck," as they said themselves. They spoke in low, earnest whispers, and Eleanor, notwithstanding her surprise, was moved even to tears. She hastily shook hands with each, and charged them to say nothing of having seen her.

"Oh, never fear, Miss Eleanor! never fear!" was the whispered response; "may the Lord be with you this mornin'—an' it's ourselves that'll miss you—but no matther—sure it's all for the best!" John then unlocked the door with as little noise as possible, and Eleanor stept out alone—*alone!* Oh, what a dreary sense of loneliness came over her as the door closed behind her, shutting her out from her childhood's home, and separating her, as it were, from

"All her youth's unconsciousness, and all her lighter cares,"

and leaving her *alone* on the threshold of a new state, without one of her family or kindred. She turned and looked up at the old house—her eye instinctively sought the windows of her mother's apartment, and a thrill of joy shot through her heart when she saw that dear mother smiling and waving a last adieu from her dressing-room window. As Eleanor kissed her hand to her mother, another face appeared for a moment at the adjoining window, and she fancied it was that of her father, whereupon she quickened her pace, in great trepidation, and almost ran till she reached the gate, where she was met by Trelawney, with Amelia and Arthur Dixon. attended by a groom, leading a horse for Eleano. Trelawney leaped from his

horse as she approached, and whispered, as he as
sisted her to mount:

"I trust, dearest Eleanor! you will never have
cause to regret this step!"

"I have no fears on that head," replied Eleanor
in a serious tone; "if I had, I would not be here
now."

"So here you are!" said Amelia gaily; "upon
my word, Nell! I owe you a grudge for taking me
out of my bed so early. I never felt more inclined
to 'slumber on' than I did this very morning. It
is really provoking to think how people *will* marry
no matter what trouble it may give their neigh
bors."

"Come, come, Emily!" said her brother, "turn
your horse, and let us be off. Don't you see
Eleanor is ready to start? We'll have Mr.
Ousely upon us, if we wait much longer, and I
give you *my* word, I'd rather meet any other man
iust now."

Larry Colgan and his wife were both out by this
time, and though they knew nothing of what was
going forward, they saw that Miss Eleanor and her
friends were equipped for a journey, and, of course,
they must wish them "all sorts of good luck!"

"I hope ye'll have a fine day!" said Larry, as
he closed the gate after them; "but, in troth, I
have my doubts about *that*."

"Why, so, Larry?" asked Sir James.

"Because the mountains are lookin' very misty this mornin', your honor, an' that's always a bad sign. I wouldn't advise ye to go very far, for the ladies, God bless them! might get a wettin' if you did. God send you fair weather at any rate!"

"Thank you, Larry!" said Eleanor; "I'm glad to have your good wishes this morning. There's something to buy a new gown for Peggy!" and she threw him a sovereign through the gate, then turned her horse to the road, and they all set off at a brisk trot.

Larry stood looking after them for a moment, then beckoned Peggy over, out of hearing of the children, who were already up and stirring. "I'll tell you what it is, Peggy!" said he, "as sure as that goold is in my hand there's somethin' goin' on. It's not for nothin' that they're all out so early this mornin'. Well! God bless Miss Eleanor any way, an' send her the heighth o' good luck wherever she goes—I'm thinkin', Peggy, it's a long journey she's settin' out on—an' none o' them with her, either!" he added musingly—"bedad, it's quare enough, so it is!" Peggy ridiculed the supposition as being "all nonsense," but Larry "knew better," he said.

A quarter of an hour's ride brought the little party to the gate of the chapel, which lay wide open for their reception. The horses were left outside on the road, "for," said Eleanor to Trelawney. in a low voice, "this is holy ground

whereon we tread; generations of sainted Christians
sleep beneath—"

"Yes! all honor is due to the lowly dead who
have died in the Lord," replied Trelawney; "I
look upon those Irish church-yards as something
really venerable—their very dust is the ashes of
saints and martyrs. But see, dearest, there is
Father O'Driscoll awaiting us at the door."

The priest extended a hand to each as they approached, and his kind, paternal smile did Eleanor's
heart good. "If one father disowns and casts me
off," said she within herself, "my heavenly Father
has provided me with another, even here on earth."

"Were you ever here before, dear Eleanor?"
asked Trelawney in a whisper, as they all followed
the priest into the sacristy.

"Yes," replied Eleanor, "this sacred edifice has
witnessed the two great events of my life; here I
was baptized only six days ago, and here, too, I
received, on last Sunday morning, at early mass,
the adorable sacrament of the altar for the first
time; ah! Trelawney, how sorry I was that you
were not there, to have had a share in that super-
human happiness."

Instead of answering directly, Trelawney uttered
an exclamation of fervent thanksgiving, which
made E'eanor start, and look inquiringly into his
face. "Nay, dear one," he said with a cloudless
smile, " you need not look surprised. Your words

have completed my happiness, for .I had no idea that you had already made your first communion. Father O'Driscoll told me of your baptism, but I have not seen him since Sunday. This is, indeed, joy."

"Thrue for you, Sir James!" said a voice near him, and the voice made all the young people start, it was so like the voice of Phil Maguire. And sure enough it *was* Phil himself, and no other, who had spoken, and there he sat on a bench in the corner, all alone, his eyes swimming in joyful tears, and his face as brimful of happiness as ever human face was. Eleanor and Trelawney smiled, and Phil nodded and smiled again, and repeated with emphasis, though in a suppressed tone : "Thrue for you, Sir James—it's you that's the lucky gentleman all out."

"I quite agree with you, Phil," whispered Trelawney close to his ear, as they passed into the sacristy, where Father O'Driscoll had preceded them. The good priest was calm and collected as usual, but the flush of joy was on his thin cheek, and his voice was somewhat tremulous. He talked some time with Eleanor and Trelawney, on the various duties of the state on which they were about to enter. When he had concluded his exhortation, he said mildly : "and now, my dear children, you can all go out into the chapel, and kneel before the altar (outside the rails) till I prepare for saying mass. I will offer up the holy sacrifice for you, before we proceed to the marriage ceremony. During mass, you will unite your intention with mine, beseeching God to prepare you for the sacrament you are about to receive, to bless your union, and to give you the graces necessary for the due fulfilment of its duties. You need not fear observation; there will be very few present this morning, besides our friend Phil, for we do not often say mass *here* on week-days."

Trelawney led Eleanor to the place appointed, while Arthur and Amelia took their seats on chairs placed for them. What high and solemn thoughts flitted through the minds of the youthful pair, as they knelt before the altar—" the altar of sacrifice" whereon was daily offered, *for them* and for all the faithful, the all-atoning sacrifice of the New Law— " the clean sacrifice" foretold by the prophet Malachy, offered up every day, from the rising to the setting of the sun, all over this habitable globe! They raised their eyes respectfully to the picture of the Crucifixion which hung over the altar—it was only a colored engraving, but viewed in connection with Catholic worship, it recalled the whole mournful scene of Calvary, and softened the Christian heart to melting tenderness. Never had Eleanor felt such sensible devotion, as when kneeling there, in the stillness of the morning, in that humble fane, with the cross before her, and

by her side him who shared her faith, and was
soon to receive her plighted vows.

When mass was over, Father O'Driscoll descended from the altar, and advanced to the rails
The boy who had served mass handed him his
breviary, the ceremony commenced, and in a few
minutes Eleanor was " a wedded wife."

" Before the altar now they stand, the bridegroom and the bride ;
And who shall paint what lovers feel, in this their hour of pride ?"

Having received the good priest's benediction,
and the congratulations of their young friends, the
bride and bridegroom both expressed their hope that
Father O'Driscoll would go over to Clareview in
the course of the afternoon to see them, as they
were to leave for England next day. Amelia
seconded the invitation, on the part of her father
and mother, and Father O'Driscoll readily consented. He went with them as far as the outer
gate, and Eleanor took the opportunity of saying
to him, in a low voice : " There is one thing more
I wish you to do for me, Father O'Driscoll, and
that as soon as you possibly can. Will you try
and prevail on Bernard O'Daly to let one of his
daughters—I do not care which—go with me to
England ?   Tell him I have the interest of his
family much at heart, and will try to advance it by
every means in my power. If he consents, the
girl must be at Clareview this evening, as we start

early to-morrow. Now mind, Father O Driscoll, I depend on you !"

" Well !" said the priest, with a friendly smile, " I know your benevolent object, Lady Trelawney, and I think I may venture to assure you that you shall have one of the girls, most probably Bridget God bless you, my child ! you have made sacrifices for the honor of His name, and be assured that he will repay you either in this life, or in that which is to come !"

" Well, I declare !" said Amelia, as they paced along together, " I don't understand why people make such a fuss about converting the Papists— this is the first time ever I was in one of their chapels, and, upon my word, Nell !—oh ! I beg a thousand pardons—Lady Trelawney ! I never felt so much like praying in all my life. It's a pity they have so much fasting, and all that kind of thing, for I really think they have the most of the piety that's going ! Heigho ! I wish they'd let people to heaven without doing penance—if they did, I'd be a Catholic to-morrow, and keep you all company ! But what in the world are you all about—examining the pebbles on the road, eh ? what a precious set of stupid geologists we have here—why, if you keep so dull and silent as you now are, people will think you are repenting already !"

Amelia kept rattling on in this way until they reached Clareview, when Eleanor received a truly cordial welcome from Mr. and Mrs. Dixon. In the course of the afternoon, Father C'Driscoll called, and was at once introduced into the drawing room.

"Well! my dear sir!" said Eleanor, the moment she saw him; "did you succeed?"

"I did!" replied the priest with a smile, as he shook hands with Mr. Dixon and his amiable wife. "You're to have Bridget, Lady Trelawney, on condition that you keep a close watch over her outgoings and incomings—those are Bernard's own words."

"Thanks, reverend sir!—the condition is one which I would, in any case, have observed. I trust Bridget will have no reason to repent of coming with me, and then there will be one less '*on the shaughran*,' as granny Mulligan would say."

"Are you not afraid of losing Mr. Ousely's friendship, Mr. Dixon?" said Father O'Driscoll, with his placid smile. "He will scarcely forgive you for receiving his truant daughter—begging her ladyship's pardon!"

"He may do for that as he likes, my dear sir," replied Mr. Dixon. "So long as my conscience does not reproach me, I care little for any man's displeasure. I think I have done him no injury in this affair," he added significantly; "there are few men who would reject such a son-in-law as he has now got. But between you and me," lowering his voice, "I don't think he's half so angry as he pretends—he daren't, for his life, offend *the saints*, you know, by conniving at his daughter's double crime of becoming a Catholic, and marrying a Catholic—you know he has his character to keep up, and must do it, let what will follow!—oh, blessed effects of the no-Popery mania!"

In the course of the evening Mr. Dixon related the scene which he had witnessed in the forge a few days before, and the company enjoyed a hearty laugh at the expense of poor McGilligan, styled by Amelia "the knight of the rueful countenance." Mr. Dixon went on to say that the honest blacksmith had been brought before the bench for the crime of burning the bible. "Fortunately," said he, "there was barely a quorum sitting, and of the three two of us were opposed to the proselytizing system, so we dismissed the case, with an admonition to the blacksmith not to burn any more bibles.'

He had scarcely done speaking when Mrs. Ousely was announced, and Eleanor hastened down stairs to have a little private talk with her mother before she entered the drawing-room. In a few minutes Sir James was sent for, and when they all three rejoined the company, Mrs. Ousely was leaning on his arm and smiling through the tears which dimmed her still beautiful eyes. When all the Dixon family had shook hands with Mrs.

Ousely, she kept looking at Father O'Driscoll, who hung back, scarcely knowing whether to come forward or not, until Eleanor led her mother towards him, saying:

"Mother! let me make you acquainted with Father O'Driscoll—now indeed *my* father!"

"I was unwilling to offer myself to your acquaintance, madam!" said the priest, with a respectful bow, "not knowing how you might be disposed to regard a Catholic priest, and especially one who has had the happiness of opening the doors of the Church to Sir James and Lady Trelawney, a heinous crime, I admit!" He smiled as he spoke these words, and Mrs. Ousely smiled too.

"Nay, my good sir," said she, "I am not quite as bad as you suppose, in that respect—I am a Protestant, indeed, and mean to continue so, but I do not go so far as to hate any one for not being a Protestant—in proof whereof, there is my hand! If my daughter thinks she will have a better chance of salvation as a Catholic, I am content!"

Very soon after the arrival of Mrs. Ousely, Eleanor was again summoned down stairs, and this time she found Phil Maguire and Bridget O'Daly.

"I am very glad to see you, Bridget! said Eleanor, pointing to a seat, "and you, too, Mr Maguire. I hope your wife is in good health."

"She can't complain, Miss Eleanor—but, bloo

alive! sure you're not Miss Eleanor now, it
seems!"

"Never mind, Mr. Maguire!" said Eleanor
blushing; "the name is not of any great consequence just now. How are your father and sisters, Bridget?"

"They're all well, I thank you, Miss—I mean,
ma'am!—my father's like a new man since"—

"Ahem!" said Phil, breaking in suddenly; "I
was wantin' to see Mister Dixon—if I could just
have a word with him."

Eleanor rang for a servant, and sent up the message to Mr. Dixon, who quickly made his appearance.

"Well, Phil! what's the matter now?—any
word from your young friend, Owen?"

"That's jist what I wanted to spake to your
honor about," said Phil, exchanging a significant
glance with Bridget, who seemed more inclined to
laugh than anything else. "I'm afeard there's
something wrong, Misther Dixon, dear, when he's
not comin'—here's a bit of a letther that came
from Galway this mornin' to poor Bernard—maybe
it'll explain the matther to our satisfaction." So
saying, he stood up and drew from behind a door,
not a letter, but Owen O'Daly himself, thin and
pale indeed, but with a bright smile on his handsome face. Mr. Dixon and Eleanor started, but
Phil was as cool as possible. "There now, your

honor," said he, " there's the letther—it's a letthei
of thanks, Misther Dixon, as full of gratitude as
an egg's full of meat !"

" Yes, Mr. Dixon !" said Owen, with deep emotion, " I am here in person to thank you for your
unhoped-for interference on my behalf, and to
assure you that neither I nor mine will ever forget
it. *Our* gratitude is not worth much, sir, but if
ever it's in the power of any of us to do anything
for you, then, sir, you'll see how grateful we can
be !"

" I believe you, Owen, my poor fellow !" said
Mr. Dixon; " I know you all better than you
think. Tell your father from me that I have a
little place in view for him, and that I'll send him
word as soon as I have all preliminaries arranged."

Owen and Phil then took their leave, after
drinking the health of the bride and bridegroom in
a couple of glasses of Kinahan's old malt. Bridget
went with them to the door, begging of Owen to
write to her very, very often, " for mind if you
don't," said she, sobbing fairly out, " I'll be home
with you very soon. Remember, Owen dear, that
it's only for the sake of being able to help my
father and all of you, that I'm going away amongst
the cold strangers—except my mistress that is to
be—and that hearing from home will be my only
comfort. Phil, be sure and tell Nanny that I'll
send her someth'ng that I know she'll like, the very

first money I get in my hands! The Lord's blessing be about her and you both!" A nd poor Bridget could scarcely get out the words; her brother could not command his voice to speak, but he squeezed her hand hard, hard, and then hurried away, Bridget calling after him : "Be sure now and let me know as soon as ever you get a letter from Cormac and Daniel, and don't forget to send me their address!"

"Never fear, Bridget!" said Phil; "among us all, you'll not be forgotten—only don't be bothering us about your presents—if I see you layin' out your money that way, you'll be in the back o' the books. D'ye mind now?—jist keep your money, *ma colleen*, for them that wants it most, an' that's what'll plase both Nanny an' me best!" Bridget said nothing, but she thought the more, and all that evening, as she sorted and packed Eleanor's clothes, which had been sent to Clareview some days before, she kept saying to herself: "That would be *one* way of showing gratitude, but it isn't *my* way!—no, indeed, if Nanny hasn't an elegant silk shawl before next summer it won't be my fault!"

When Mrs. Ousely came to take leave of her daughter, she was not half so much agitated as might be supposed, and when Eleanor clung to her neck in an agony of weeping, she softly whispered "Be comforted, my daughter! we shall soon meet

again—believe me we shall!" Eleanor wept no
more.

On the following day, when the new-married
pair reached Galway, accompanied by Amelia
Dixon, who was to spend the winter with them in
Somersetshire, the first person they saw, on reaching the hotel where they were to await the sailing
of the packet, was Mr. Ousely, whip in hand, who
appeared to take no notice of them as they passed
in, but they were scarcely seated in the parlor
when he bolted in, and nodding to Trelawney,
went straight up to where Eleanor was sitting, and
planted himself right before her. Trelawney drew
near, fearing that he meant to strike her, and
Eleanor, pale with apprehension, could only falter
out :

"Father! you here? I did not expect—
what"—

"I'm here to see *you* off, you ungrateful, unduti-
ful hussy!—d——n it, Nell! how could you think
of treating your father so?—there—that'll do—I
forgive you, Papist and all as you are. Give me
your hand, Trelawney!—hang it! I'm not as bad
as I seem—*my bark is worse than my bite*, as the
adage says. The old woman and myself are going
over to see you soon. Let them say what they
like, Nell! you're my daughter still. Now, Amelia, I'm not so bad, after all, you see!"

"I'm very glad to hear you say so, Mr. Ousely!" replied Amelia drily; "you ought to know best."

"My dear father!" said Eleanor, taking both his hands in hers; "how happy you have made me by this most unexpected kindness! Your presence now is like balm to my heart, for the thought of having incurred your displeasure would have embittered every moment of my life!—May I hope," she added with her sweetest smile, "that you will extend your forgiveness to my partner in guilt?"

"Allow me to support the prayer of the petition, Mr. Ousely!" said Sir James, coming forward with outstretched hand. "You *ought* to forgive me, my dear sir, for you must admit that if I robbed you of a daughter I have given you a son-in-law."

Ousely looked at the offered hand for a moment, as though he were undecided, then suddenly taking hold of it, he gave it a hearty shake:

"D——n it, I suppose I *must* give in. I *had* my mind made up to shoot you the first opportunity, but now I think I'll go home and have a chance at Dixon—eh, Amelia?"

"Oh, pray don't, sir," said Amelia, with mock seriousness—"pray, don't shoot papa—he'll never do the like again!"

"I believe you," said Ousely, with a laugh; "he can never offend *me*, at least, in the same way. Come along, Trelawney, and let us see about

having a lunch—after that there are some pecu-
niary affairs to be settled before you go—Nell
Ousely must not go to her new home a beggar after all that's past and gone!"

CHAPTER XVII.

' Last scene of all that ends this strange, eventful history.'
SHAKSPEARE

"Here, too, dwells simple truth ; plain innocence ;
Unsullied beauty ; sound, unbroken youth,
Patient of labor ; with a little pleased ;
Health ever blooming ; unambitious toil."
THOMSON'S Seasons.

A YEAR had passed away, after the events recorded in our last chapter, and it had, as usual, brought many changes ; for there is no year that rolls away into the far depths of eternity, without producing some revolution, or effecting some change " in the affairs of men." Bernard O'Daly had moved into the house so kindly given him by Mr. Dixon, and through the kindness of Phil and Nanny, and a few other neighbors who were still in a condition to give a little help, it was soon provided with the little plenishing which the family required. Owen had regular employment at Clareview, where he was engaged to assist the gardener all the year round, and Mr. Dixon gave him a small plot of ground, on very moderate terms, which he cultivated before and after hours, and thus raised as much vegetables as helped to

support the family, now much reduced in numbers.
Mrs. Ousely, unknown to her husband, gave Kathleen so much sewing to do, that it kept both her
and Evaleen employed during the intervals of their
household duties. Peace had again settled down
on this long-afflicted family, and the sorrows
of the past were already assuming a softened, hazy
character in their remembrance. Bridget was still
with Lady Trelawney, who had fulfilled to the
letter all the promises she had made in her favor,
and regularly every quarter, there came a few
pounds for Bernard, together with sundry presents
for Owen and the girls, which, with their own
earning, enabled them to resume somewhat of their
former respectability of appearance. Phil and
Nanny were still jogging on "thegither," as kind
and as eccentric as ever. Bridget O'Daly had not
forgotten her promise in regard to Nanny, and, on
every fine Sunday or holyday, the good woman was
seen trudging along to the parish Chapel, with a rich
shawl of crimson silk covering her broad shoul
ders; after a little there came a handsome merino
dress, and lastly, a fine Tuscan bonnet, with a great
plenty of broad, rich ribbon, and when Nanny was
attired in all this finery, it was her pride to tell
the neighbor women who gathered round her in
the chapel-yard after mass: "They're all presents
from Bridget O'Daly, and came all the way from
England beyant! Isn't it past the common, the

goodness of that girl ? See what a mint o' money
she must have laid out on them !—an' jist look at
our Phil, yondher—well ! it was Bridget sent him
that beautiful silk handkecher he has on his neck.'
These announcements were heard with all due ad
miration, bu: most of the women wound up their
praise of the objects themselves, and the kindness
of " them that sent them," with, " but sure she
done nothing but what she had a right to do—it's
yourself an' Phil that was the good, kind friends
to them all, when they wanted them badly, poor
things !" " Well ! of coorse," would Nanny say,
" we did what we could, an' maybe a little more,
too—*na bocklish*——but still, it's an ould sayin',
you know, that *eaten bread's soon forgotten ;* but
it's not so with the O'Dalys, the creatures ! they
have the ould dacency in them yet, an' the piety,
and the goodness—"

" Signs on them, Mrs. Maguire !" said old Judy ;
" sure they're gettin' over their throuble bravely,
thanks be to God for it, an' they're beginnin' to do
well again !"

" Betther an' betther may they do, then !" said
Peggy Colgan, who was one of the group of listeners. " I know the misthress up at the Hall
thinks a power of them, an' keeps the girls
constantly in work. When she was over seein'
Lady Trelawney last summer—you know she
spent a month with her——' she took a great likin'

to Bridget. It seems there's a young heir born, an' Bridget's got to be nurse, an' she has a great advance in her wages—so you see, *when luck comes it comes jumpin'.*"

"But do you tell me, Peggy, that there's a son come home?" asked Nanny, in surprise. "Why, I didn't hear a word of it."

"Well, it's thrue for all that," replied Peggy. "The misthress herself tould me when I was openin' the gate for her this mornin, an' her goin' out to church."

"Dear me, then," said Nanny, "I must go an' tell Phil an' the rest o' them," and away she bustled, brimful of the glad tidings she had to communicate. First she told it to Phil, who rubbed his hands, and cried: "Blood alive! Nanny, that's great news—I declare I'm as glad as ever I was in all my life; an' you tell me that Miss Eleanor—pooh, bad cess to this memory o' mine —I mane her ladyship is well—"

"'Deed an' I didn't tell you any such thing," retorted Nanny, "for I never thought of askin'; but I suppose she's well, or Peggy'd have said so." Then she hurried off in search of the O'Dalys, whom she found assembled at Honora's grave. Nanny had too much natural delicacy to open her news-bag in such a place or at such a time, so she knelt with the others, and offered up a fervent prayer for the repose of the soul of her ancient

friend. It was not till she brought them all back
to where Phil was standing, that she told them
what had happened, and then they were so "overjoyed," as they said themselves, that they forgot
all about their recent sadness, and the tears were
quickly wiped away. "Musha! the Lord be
praised!" cried Bernard; "I'm a poor man this
blessed day, an' I'd rather hear that news than if I
got a purse of goold—indeed, I could scarce hear
anything that 'id plase me better, barrin' it was
news from Cormac and Daniel."

"Well! come home now an' have your dinner
with us," said Phil, "an' we can talk over all the
news." To this Kathleen made some objection,
but Nanny laid hold of her arm, and taking Eve
leen by the hand, said :

"None of your nonsense now—I'm sure we're
no sthrangers, that you'd be makin' excuses that
way. There now, Phil, do you bring Bernard and
Owen with you, I have *my* share!" so off she
marched with her two laughing prisoners, while
Phil brought up the rear with Bernard and Owen.
Before they left the chapel-yard, however, they
made it their business to see Father O'Driscoll
and tell him the good news from England.

"I am much obliged to you all," said he, with
his accustomed smile, "for coming to make me a
sharer in your joy but I heard the news yesterday
and I can tell you rurther that the young heir of

Trelawney House is called Thomas Harrington,
first in honor of the saint on whose day he was
born, and next in compliment to Mr. Ousely, who,
I am told, is quite elated." He then shook hands
with Eveleen, and hoped she was still a good, du-
tiful girl;—her father answered for her that he
couldn't complain of poor Eveleen—she was always a good, obedient child.

"Any word from America yet, Bernard ?" inquired Father O'Driscoll.

"Not since that last letter that I showed your
reverence—I'm beginnin' to be uneasy about the
boys, for you know they said in that letter that
they'd soon write again."

"Oh, but you must not be uneasy, Bernard !—
your boys are in good keeping—God will watch
over them and you too—make yourself easy on
that head, and you will soon see that there is no
cause for apprehension !—I must now bid you all
good bye, or Nancy Breen will raise a storm about
my ears if I let my breakfast be spoiled. God
bless you all !"

The chapel was already far behind, and our little
party trudged merrily along, while "talk of various kinds beguiled the road." They had got
about half way to Phil Maguire's house, when Andrew McGilligan passed them by, his books, as
usual, under his arm, and his broad-brimmed hat
pulled down over his brows.

"Hillo, Andrew!" cried Phil, winking at Ber
nard, "what's your hurry, man alive ?—can't you
take time to give us a verse or too—*do*, Andy
dear, we're all poor Papists here, *thirsting for the
word*," and Phil imitated the nasal twang of the
Conventicle to such perfection that no one could
help laughing, but Andrew walked on, 'fast and
faster,' and never once turned his head.

"Begorra," said Phil, "he's afeard of bein'
turned into a pillar of salt like Lot's wife! Och,
then, Andy *ahagur*, but it's althered times with
you, honey, when you'd pass us by with the could
shoulder. Dear, oh dear! what's the world comin'
to, at all ?" Still Andy kept never minding.

"But, that's thrue, Andy, did you hear the
news ?" cried Phil, raising his voice as the distance
between them increased. Andrew was seen to
slacken his pace, but still he never looked behind.

"I say, *acushla !*" shouted his persevering inter-
rogator. "Did you hear what happened your
friend O'Hagarty the other day ?"

"No," said Andrew, comin' to a full stop, and
facing round; "I trust no evil has befallen him ?"

"Evil enough," replied Phil; "he was taken
sick afther a surfeit of dhrinking, and kicked the
bucket."

"What !" cried Andrew, opening his eyes wide;
'you don't mean to say that he died ?"

"That's just what I *do* mane to say !" rejoined

Phil; "but that wasn't the worst of it, Andy
dear—it was't bad enough for him to die, but he
thought fit to call for the priest when he found
himself goin'."

"Poor man   poor man!" quoth Andy; "he
must have lost his senses!"

"You mane to say, he began to find them. But
you needn't be so shocked, Andy, the devil had
too fast a grip of him to let him slip that way,
afther him sarvin' him so long.   There was a body
guard of your 'dearly beloved brethren' about
him to see that no priest came, an' the more the
unfortunate man cried out for a priest the higher
they raised their voices, telling him to 'hope in
God,' an' to 'believe on the Lord Jesus,' an' that
that was all he had to do.   Och, the curse o' God
villains! they done their duty well, an' kept quotin'
Scripture to the man that was only answerin'
them with oaths an' curses, till the poor, miserable
soul left the body an' went to its account."

The last part of the discourse was lost on Andy,
who had quickly scampered out of hearing.   Phil
himself, and those who were with him, were so
shocked by the terrible picture thus presented to
their minds, that for some time they walked on in
silence.

"May the Lord save us all from an ill end!"
said Nanny at length.   "It's enough to frighten the

lin in one to think of such a death as that. Och,
och, but they're well guided that God guides!"

" You may say that, Nanny dear!" said Bernard,
with a heavy sigh; " I had heard before now that
poor O'Hagarty was dead, God pardon him his
sins! but I didn't hear anything of how or when it
happened. To tell you the thruth, I was sorry
when I heard it, for, bad as the crature was, he
wouldn't sware agin Owen in the wrong. Ah,
then, Phil dear! how did it happen that he died
without the clargy?—was there no Christian within
hearin' that 'id go for the priest?"

" There was," said Phil, " one or two Catholic
sarvints in the house, an' one of them, hearin' the
poor man pladin' with the black-livered Jumpers
an' Scripture-readers to send for a priest, went
straight to the priest's house, but as ill luck 'ud have
it, he was out on a sick call, an' the girl darn't
take time to go to the other end of the town, where
there was another—at any rate, by the time she
got back, the poor man was at the last gasp, an'
they say it was pitiful to see him. The very last
words he said were: ' Oh Lord! oh Lord! the
shadow of the cross won't rest on *my* grave!—oh
misery!—I'm lost!—I'm lost!' An' so he died.
When the long-nosed, black-faced genthry were
quite sure that he was gone, an' that there was no
more danger of his dyin' a Catholic, they went off
an' left him to the people o' the house to get him

buried in the best way they could, only tellin' them
that he was to go to the Prodestan' buryin'-ground.
They say there was a great show-off of Scripture-
readers, an' Jumpers, an' all such riff-raff at his
funeral."

"What a lamentable death!" said Owen. "The
poor dying sinner pronounced his own condemna-
tion, as very often happens. I suppose you all
heard of that other priest who came back a few
weeks ago here in Connemara."

"No," said Phil, "we didn't hear anything of it.
Where did you see it?"

"Why, in a Dublin newspaper that Father
O'Driscoll lent me. It seems that the Protestant
bishop was going to give confirmation, and the
minister requested this priest—I forget his name—
to prepare for being confirmed on a certain day.
It's likely that he had been thinking about the state
of his soul before that, for all at once he took a
notion and went to the real bishop, who was also
in town at the time, and humbled himself before
him, begging to be received back into the Church,
and that he'd do anything at all the bishop might
choose to lay upon him as penance for the crying
scandal he had given the faithful. The bishop, of
course, consented, and the poor priest made a
public recantation, and tried to address the people
present in the church, but couldn't go on, he was
so deeply affected, between shame and sorrow."

" Ah !" said Phil, " but he must have been a
very different man from poor O'Hagarty—I suppose he had only been a short time out o' the
church, and hadn't led sich a bad life, or the bishop
wouldn't have received him so easily !"

They had now reached the house, and were
agreeably surprised to find the dinner almost
ready, for Katty Boyce, seeing them linger so long,
had quietly slipped away, and set about cooking
the dinner, having overheard the invitation given to
the O'Daly family, " an'," said she, ." I knew very
well that it 'id be very late when yez 'id get home, an'
that it wouldn't be any affront to find the dinner
near ready. By good luck, I was in here this
mornin' awhile, an' seen what the misthress laid
out for the dinner."

" By the laws, Katty, it was a lucky thought !"
said Phil, " for we're all starvin' with hunger."

" It was well you knew where to find the kay,"
observed Nanny, as she threw off her cloak (it
being mid-winter), and hastened to assist Katty.

"So you don't go any more to the soup-house,
Katty ?" said Owen, slily.

" Ooh, musha, but it's myself that does not,"
replied Katty, as she wiped the perspiration from
her face, with the corner of her clean check apron ;
" since I fell in with these people here, I never
knew what want was, nor the children neither,
glory be to God !—sure the boys goes every day

to school to *our own* school-house, an' afther school
they come over here, an' do little turns for Misther
Maguire, an' sure the misthress an' him keeps
clothes on them, an' I declare to you, Owen, we're
as happy an' contented as if we were in the king's
palace!" And the cheerful, ruddy countenance o'
poor Katty was an unmistakable proof that she
spoke the truth.

About two days after, Eveleen was sent to the
post-office in Killany to see "if there was any
luck," and her timid question of "Is there anything
for Bernard O'Daly?" was answered by the ready
response: "Yes, my little girl, there's something
for you to-day. Here's an American letter, and a
money-letter, too. You haven't your journey for
nothing this time."

"How much is on it, if you please, sir?" said
Eveleen, as she put the precious letter in her bosom. "My father only sent a shilling with me."

"Well! there's four pence more on it," said the
postmaster, "but you can give it to me some other
time."

Eveleen ran the whole way home, and when she
got near the house, she saw Kathleen standing on
the ditch, watching for her.

"Well, Eveleen, what news have you?"

"Good news, Kathleen—a great big letter with
Cormac's own hand-writing on the back and Mr.
Brown says it's a money-letter."

Their father met them at the door, and before he had time to speak, Eveleen put the letter into his hand, having first kissed it over and over. Agitated by his excessive joy, the old man could scarcely keep his feet, and his daughters led him to a seat. Then, when he did attempt to open the letter, his hands trembled so that he could not succeed in breaking the seal. Handing it to Kathleen, he said : "There, Kauth, darling! do you read it— joy is a'most as powerful as grief, children!—I wish poor Owen was here at the openin' of it, but sure we can't have *every* thing we wish—the Lord make us truly thankful for what He sends us! Go on, Kathleen dear, let us hear what's in it."

The letter was written one half by Cormac, and the other by Daniel. It spoke of many trials and hardships, " all past and gone," as the brothers gratefully said, and of sudden prosperity pouring in upon them when they least expected it ; and as a proof that they did not forget the condition in which they had left " the loved ones at home," they inclosed a draft for *one hundred dollars*, being *twenty pounds* sterling. (They had sent five pounds before, which had been laid out on a suit of clothes for Bernard, and another for Owen.) They both assured their father that he need not want for any comfort, for that they were both in good situations, Cormac as steward of a steamboat, and Daniel as clerk in a store, and that he might depend on having a remittance from them every three or four months. "We would send for you all, my dear father," wrote Cormac, "were it not that we thought it would be more agreeable to you to spend the evening of your life in the place where your youth was passed, and where our dear mother lies. We knew that to take you away from poor old Ireland, would be like the parting of soul and body, and so we made up our minds to let you remain there, with Owen and our dear sisters. We are rejoiced to hear that Bridget is so well situated, and that she still shows herself what she always was, a good and affectionate daughter. Give our kind love to Phil and Nanny Maguire, and to all inquiring friends, not forgetting granny Mulligan, (whom Owen forgot to mention in his last letter.) Give our best respects to Father O'Driscoll, and tell him that we never forgot his parting words, and with God's help never will. There's a great deal of noise here about the proselytizing in Connemara, and it often makes us laugh (though it's provoking enough, too,) to hear of *the great reformation* going on there. It would be a real farce to us, who know how matters really stand, were it not connected with the fearful sufferings of our people, who have not only famine and pestilence to contend with, but also this so-called *Reformation*, perhaps the greatest plague of all! When you write, tell us all about it, but I suppose it has nearly died a natural death by this time, seeing that the famine is well nigh over. Tell Father O'Driscoll that either Daniel or I will write to him very soon. Pray for us, my dear father and my good sisters, you who rest quietly in the old ark of peace at home, while we are tossed about on the restless waters of this great commercial world. Pray for us that we be not 'led into temptation. God's blessing be with you all!"

Whilst Kathleen was reading the letter, her father sat with his hands clasped on the top of his stick, and his eyes fixed on his daughter's face, while the big tears rolled unheeded down his cheeks. When it was all read, from date to signature, the old man drew a long breath.

"Well, thanks be to God!" said he, "they're doin' their own share, any way, for us. Sure enough, I'm the happiest father on Irish ground, an' I don't desarve the tithe of the good gifts the Lord is sendin' me!—och! och! children dear, if your poor mother had only lived to see this day, it 'id banish the cowld grief from her heart, an' make her eyes shine as bright as they did when she was a purty, fair-haired *colleen*, years an' years ago. But then isn't it sinful to wish her back on this miserable earth, where joy only comes in little weeny blinks now an' then—och, what's the happiness that *we* have here, to the never-endin' glory an' happiness that she now enjoys? for if she's not happy," he added, in a sort of soliloquising tone, "then God help the world !"

"But, father dear!" said Kathleen, smiling through her tears, "you're not asking to see the draft."

"Why, then, it's thrue for you, Kauth! I was forgettin' all about it! Give it here, Eveleen, my daughter." The little girl had been examining the precious document, as a sort of curiosity in its way, and she said, as she handed it to her father : " Well! isn't it curious how that little bit of paper can be worth twenty pounds !"

When Bernard had carefully inspected the draft, with the help of his spectacles, he pulled out an old leather pocket-book, which might have been in the family " since the wars of Ireland," and in it he placed both the letter and its inclosure, the former being, if anything, the more valuable of the two, at least in Bernard's estimation. When Owen came home in the evening, Eveleen met him at the door with the good news, and he had scarcely crossed the threshold, when his father handed him the letter. A flush of joy crimsoned Owen's fine features, as he read the hope-inspiring words penned by those brothers, so dearly loved, and so " far, far away." Cormac had mentioned, in the earlier part of his letter, that he and Daniel would be more than glad to send for Owen, but that they supposed he would not think of leaving home, so long as God spared their dear father.

"He is right!" said Owen, with generous warmth; "it's the least that the father of three sons should have *one* of them to lay the sod over him when it pleases God to take him to himself."

"God bless you, Owen! God bless you, my son!" said his father, his eyes filling with tears. "I'm sorry to have to keep you from where you might have a chance of risin' in the world!"

"You needn't be a bit sorry, then, father, on that account," replied the young man warmly; "for I can tell you it's proud and happy I feel to be the one that God has pitched upon to be near you, and to comfort you in your old age, especially as it's an honor I couldn't expect, being the youngest."

The old man smiled and shook his head, and told Kathleen to make haste with the supper, "for sure the poor boy must be in need of it by this time."

In the course of the evening, Bernard and his children held a consultation as to what was to be done with their newly-acquired wealth. After some deliberation, the old man said:

"That's the first thing to be done at any rate, so, plase God! I'll take a walk over there some time to-morrow."

Whatever the proposal was, it was quite agreea

ble to the young people, and so the matter rested
for that night. Never had the Rosary been said
with more fervor than it was on that night, for as
Kathleen said :

"We offered up many a prayer to the Mother of
God when we were in sore, sore need, and it's the
least we can do to thank her now, when she has
obtained so many blessings for us, and brought us
safe through all our trouble."

The prayers once over, the happy family retired
to rest, and their slumbers were calm and sound,
for theirs was precisely the condition which attracts
"tired nature's sweet restorer—balmy sleep," who

"————— Like the world, his ready visit pays
Where fortune smiles."

Early next day Bernard set out on the well
breaten track that led across the fields to Phil Maguire's. On reaching the comfortable old homestead he found Nanny alone in the house, and she
hard at work whitewashing her kitchen.

"God bless the work, Nanny !" said Bernard, as
he entered ; "where's the good man from you this
mornin' ?"

"Why, then, Bernard O'Daly, is this yourself?"
cried Nanny, giving her brush a shake over the
pail. "The sorra one o' me knows where Phil is,
barrin' he's out in the byre, fotherin' the cattle.
But, sure, it's newens to see you out so early !"

"Thrue for you, Nanny, an' maybe I wouldn't be so early afoot *this* mornin', only that I have a little business with Phil. I'll just step out myself, an' see if he's about the house."

Nanny's curiosity was fairly excited, and her mind was, at least, as busy as her hands, until Bernard came back in a few minutes with Phil.

"Come an' take an air o' the fire, Bernard," said Phil; "it's freezin' hard." So saying, he began to rake out the hot *greeshaugh*, while Bernard, on the other hand, was taking from his old pocket-book the highly-prized American letter, which was very quickly discovered by Nanny's keen eye.

"Eh!—what's that, Bernard?—have you got a letther from the boys?"

"Deed an' I have, Nanny, an' that's what brought me over this mornin'. There's the letther, Phil, an' see what was in it." He handed the letter and the draft to Phil.

"How much is in it, Phil?" said Nanny, suddenly dropping her brush, and sitting down on a *creepy* beside her husband.

"Blood alive!" said Phil, after looking at the draft. "Twenty pounds—not a penny less!—By the laws, Bernard, you're a rich man this mornin'."

"Twenty pounds!" repeated Nanny in amazement. "Why, Lord bless me, Bernard, what will you do with all that money?"

"Oh, I'll find use for it, never fear!" said the

old man, with a smile. "But go on an' read the
letther, Phil, an' I'm sure both of you'll say that I
have the best sons that ever stepped in shoe
leather—God reward them for it!"

When Phil had got through the letter, Bernard
said very quietly:

"And now, Phil, I've something to say to you—
there's a part o' this money that belongs to you!"

"To *me!*" cried Phil, staring at him in astonish
ment; "why, how would any of it belong to me,
in the name o' goodness?"

"What in the world do you mane, Bernard?"
exclaimed Nanny.

"Why, then, I declare you're the simplest pair
in the world wide," said Bernard, "or you'd know
very well what I mane—I want to make you some
allowance for all the time that myself an' the
children were on your floor, an' eatin' your bread.
Myself and Owen jist settled it atween us last
night, that I'd come over this mornin' with the
dhraft, an' let you take whatever you like out
of it!"

"Yourself an' Owen might have employed yourselves betther than settlin' any such thing," returned
Phil testily, "an' if I had a known what was
bringin' you here, I'm blest and happy but I'd have
given you the door this morning, cowld as it is.
If it wasn't your own four bores that's in it, Bernard O'Daly! I vow to God, I'd never change

words with you, afther makin' me such an offer!—
put *that* in your pipe and smoke it!"

Nanny's cupidity was at first strongly excited by
Bernard's proposal, but on hearing her husband's
burst of generous indignation, her own better nature triumphed, and she said:

"Hut, tut! Bernard! didn't you know very
well that what we done was done for God's sake,
an' for the sake of ould friendship?"

"I know, Nanny, I know that very well, but
still an' all, it's only fair that when God sends it to
me, I'd make you some return."

"Now, I'll tell you what it is, Bernard!" said
Phil, laying his hand on his knee, "if ever I hear
you spake of sich a thing again, I'll never open my
lips to you while there's breath in my body.
Nanny! rise up an' get us that black bottle that's
in the cupboard there—this poor foolish old man
'ill be the better of a glass this frosty mornin'
afther his walk."

"Thank you all the same, Phil, but I'd rather
not take anything. I'm jist on my step down to
Father O'Driscoll to show him the letther."

"Bad cess to the foot you'll stir out o' this, till
you take something to warm you—make haste,
Nanny." So the black bottle was brought, and
the quarrel was made up, but not until Bernard
had to promise that he would never again offend in
a similar way, and then Bernard set out with re

newed spirits for Father O'Driscoll's. He found
the good priest busy giving instructions to no less
than four of the poor perverts, who having got
work from one farmer and another, were no longer
in need of *the soup*, and came to seek forgiveness
from their long-deserted pastor, and a reconcilia-
tion with that old, venerable Church, which had,
they trusted, sent generations of their kindred to
heaven.   Bernard was leaving the room when he
perceived what was going on, but Father O'Dris
coll called him back, observing that the penitents
whom he saw there were quite willing that their
return to the 'one fold' should be made public, in
order to make satisfaction for the scandal they had
given.

   " But, indeed, indeed, your reverence," said one,
" it wasn't our faut. I know very well that we
ought to die of hunger sooner than run the risk of
losin' our souls, an' maybe if we had only our
selves, Father O'Driscoll, we *might* hould out to
the last, but, ochone! when a man sees the wife
an' the little ones faintin' and dyin' with hunger
before his eyes, an' himself worse than any o'
them—when the food is neither to be had for askin'
nor earnin'—och, sure, it's hard to stand it—sure
it is, your reverence, especially with the divil whis-
perin' at one's elbow, 'Go to the soup-shop—
there's plenty there—if you let *them* die it's your
wn faut!' Nobody knows, your reverence, --

sept God alone, how hard it is to stand that temp
tation."

"I know it, Thady!" said the priest, soothingly.
"I know it well, my poor fellow—tne tempter
come to you in your sorest need, armed with money,
food and clothes, while we have nothing to give
but our prayers and our sympathy. Ah! it is
terrible, terrible, the struggle that you have to
maintain between faith and famine!"

Bernard, seeing that there was no immediate
prospect of having Father O'Driscoll alone, went
forward and gave him the letter, saying he should call
for it next day. When he got home, the first object
that presented itself was granny Mulligan's big
bag lying on the table, and the next was its owner
herself, seated in her usual place by the "ingle-
side." She was smoking away from a short cutty
pipe, and, at the same time, giving directions to
Eveleen, who was trying her hand at a potatoe
cake, on the table near her.

"Why, granny Mulligan, in the world wide is
this you?" said Bernard. "You're jist the very
woman I'm glad to see. We thought you were
down about the Lake side, somewhere."

"So I was, Bernard!" replied the imperturbable
old woman; "but I heard a flyin' report last night
that you got a letther from the boys, so I cut my
stick from Neddy Breen's, where I was, an' made
the best my way up, this mornin', to see if it

was thrue. You needn't throuble yourself tellin
me, now, for the girls tould me all about it."

" But we didn't tell her about what Bridget sent
for herself, father !" said Eveleen ; "we left *that*
for you.".

"Ah, now, do you tell me that Bridget sent me
something all the way from England ?" said granny, taking the pipe from her mouth.

"She sent you this !" replied Bernard, going into
the room, and returning with a very handsome
rosary of cocoanut beads, linked with silver, and
having a pendent crucifix of the same metal,
Granny Mulligan's eyes filled with tears, as she
took the beads, and carefully examined their various
beauties, not one of which escaped her observation.

"So you see, Bridget doesn't forget poor granny
Mulligan," said she at length, as she wiped away
a trickling tear. "An' Kathleen tells me, too, that
Cormac and Daniel both sent their love to me."

" An' more shame for them if they didn't !"
observed Bernard.

"Hut, tut ! Bernard, don't say *that !*— the
world's gettin' so cowld, an' the people are all so
taken up with themselves, that it does an ould body's
heart good to see a spark of kindness or gratitude
in the young people risin' up. But sure there's
nothing in these childhren o' yours but good-nature
an' the heighth o' friendship, an' they had *that* in
them since they were weeny things runnin' around

Well, indeed, I'm proud an' thankful that they all remember poor culd granny."

In the course of the evening, Kathleen told granny that she must give over rambling, and spend the remainder of her days with them; "for now, thanks be to God! we have the manes of keepin' you comfortable!"

For a moment the old woman was silent; her lips moved as though she were talking to herself, and then there came a big tear trickling slowly from either eye; Eveleen put her arm coaxingly round her neck, and said: "Ah *do!* granny—*do* come and stay with us altogether!"

"Well! I b'lieve I must give in, childhren!" said granny, all of a sudden. "I never thought to see the day when I'd agree to settle myselt down, an' never go out again among the ould cronies, that gave me a warm corner everywhere I went; but Parson Hendhersou tould me, no later nor yesterday, when he met me on the road, that he'd have me taken up for a vagrant, if I went about beggin' any more, so, for fear he'd keep his word, I think it's best for me to conteut myself here among the good Christians that'll make me welcome, an' not put myself in the power o' them black-hearted villains o' the world, that wouldn't desire betther than to get an excuse for tormentin a poor ould Papist like me. Phil Maguire an' Nanny made me welcome to go an' live with

them, so I can stay part o' the time with them, an
the other part with you. Father O'Driscoll, the
Lord's blessin' on him ! always gives me the price o'
the tobaccy, an' always will, he says, while he's
in the parish ! So, in the name of God, Kathleen
dear, I'll do what you bid me, for I *know* I'm wel
come, an' that I'm with *my own* while I'm with
you !"

"That's right, granny, that's right !" said Bernard; "so mind this is your home for the time to
come. Sorra poorhouse you'll go to while *we* have
a shelther for you."

"But, surely, granny !" said Owen, who dearly
loved a joke, "surely, you didn't ask charity trom
Henderson ?"

"Is it *me* ask charity from him !" exclaimed the
old woman in a tone of the most supreme con
tempt. "Do you think I'd be such a fool ? oh,
then, indeed, I didn't, an' it's what I said to him
when he taxed me with bein' a beggar : 'Did I ever
ax *you* for anything ?'—says I to him. 'Faix I
didn't, bekase I knew very well that I wouldn't get
it, barrin' it 'id be a thract or a testament, Misther
Hendherson, an them's very poor comfort for hungry bellies !' With that he rise his whip to me,
an' bid me be off, for a roublesome old Romanist —
I think that was the word !"

So now we have settled granny Mulligan "in
pace an' quiteness," as she said herself, with the

O'Daly family—all "well and doing well;' ditto Phil Maguire, and his close-fisted, yet charitable helpmate; Sir James and Lady Trelawney safely moored under shelter of the old rock, or in other words happily embarked in the stout old ship—of which Peter is the helmsman, and Our Lord himself the pilot; Father O'Driscoll is still breasting the torrent of persecution, and waging successful warfare, in his own quiet way, against the hydra-headed monster of Proselytism; we have shown poor Andrew McGilligan foiled on every hand in his attempts to spread what he calls "*Gospel truth*," and relaxing his efforts in sullen despair. It only remains to say a word of the Dixon family. Amelia, during her stay with Lady Trelawney, renewed her acquaintance with Lieutenant Gray, who with his friend Captain Hampton, was then stationed in the neighborhood of Trelawney House. She very soon cured the young officer (who was not without a certain amount of good sense) or the lisping dandyism which he had allowed himself to contract, and as he had a small property in addition to his pay, they managed, as Amelia wrote to her mother: "just to keep their heads decently above water, and let people see that they were somebody!" Mr. Dixon and Mr. Ousely very soon made up the quarrel, and went over to England together soon after the birth of Eleanor's son, to visit their respective daughters and son-in-laws

on which occasion Ousely gave great offence to
Mrs. Hampton, by forswearing all future connection with the Jumpers and Proselytizers, and consigning them to warm quarters in the other world.
Mrs. Ousely and Mrs. Dixon accompanied their
liege lords, and they were all so charmed with
their visit that they could scarce make up their
minds to return home. This was especially the
case with Mrs. Ousely, who, unlike her friend, had
now no tie to bind her to Ireland. Finally there
was a compromise effected, to the effect that Trelawney and Eleanor should spend part of each year
in Ireland, Mr. Ousely declaring that, with all its
poverty and Romanism, he'd rather, a d——d sight,
live in Ireland than in England."

"Why, my dear father!" said Eleanor with her
arch smile, "I don't wonder at your preferring
Ireland and 'the old house at home,' where you
have the full blaze of those New Lights, which
must, surely, have spread their radiance far and
wide by this time, seeing that they were burning
so brightly when I left, now better than a year
ago!"

"Blast them for New Lights!" cried her father
pettishly; "they're nothing but confounded *will-o-
the-wisps*, as I can tell to my cost. I don't mean to
say that I've any greater love for Romanism than
I had, save and except this Papist daughter of
mine and her better half—but I've got my eye

opened of late to the goings on of these same New
Lights, and I say they're doing no good for either
king, country, or religion."

"Never mind, Ousely!" said Dixon, tapping him
on the shoulder, "they'll soon burn out—you and
I may live to see the good old times back again—
by George! there's more life, and light, and heat,
in what is facetiously termed 'the darkness of the
Irish people,' than in this unnatural flare kindled
by the Proselytizers!"

"As far as Eleanor and myself are concerned,"
said Trelawney, "I can assure you that we owe
our conversion solely to these same New Lights,
so that we, at least, are much indebted to them."

CONCLUSION.

By way of introducing some observations which I mean to make on the proselytizing system in Ireland, I think I cannot do better than lay the following extracts before the reader, with the single remark that they are all from Protestant writers whose words I give *verbatim*.

"There is not in the world a more modest race of women than the Irish; a remark which equally applies to all ranks and classes among them. . . . The Irish are a most obliging, kind-hearted, and hospitable people. In all these qualities they are unequalled by any other nation in Europe. To have an opportunity of obliging, or showing attention to a stranger, affords an Irishman a pleasure of the highest order. . . . The Irish are a nation of practical philanthropists; they rejoice in the happiness of others. They are happy if they can only promote the happiness of strangers. One might travel from one extremity of the Island to another, without having cause to complain of a cold look, an unkind word, or an ungenerous action . . . As regards hospitality, again, it is known that the Irish have always been proverbial. They will

share their last meal with you, and be miserable
if you refuse to participate of it. . . .   Even the
poor peasant, who has only his one meal a-day,
and that consisting of potatoes, will cheerfully divide it with any poor creature who chances to pass
his door. . . .   How unlike the poor of this country!   There is little sympathy towards each other
among them. *We are, as compared with the Irish,
an unfeeling and selfish people.*"—IMPRESSIONS OF
IRELAND AND THE IRISH.

"Everybody knows that a stranger could travel
in the worst of times, and in the worst districts,
at all hours of the day or night, with a charmed
life, and, in fact, never be insulted or molested."—
STARK'S TOUR IN IRELAND.

"As regards the women of Ireland, their native
modesty cannot fail to attract the observation of
any stranger."   "From the morning on which I
had visited the great model National School, in
Marlborough Street, Dublin, to the hour of my
arrival at Galway, I had remarked, in the Irish
female countenance, an innate or native modesty,
more clearly legible than it has ever been my fortune to read in journeying through any other country on the globe. . . .   I am convinced that no
man of ordinary observation can have travelled, or
can now travel, through Ireland, wi.hout corrobo-
rating the fact.

"But I have lived long enough to know that

outward appearance cannot always be trusted, and, accordingly, wherever I went, I made inquiries, the result of which was not only to confirm, *but to over-confirm*, my own observation; indeed, from the Resident Commissioner of the Board of National Education in the metropolis, down to the governors of jails and masters of the remotest workhouses, I received statements of the chastity of the Irishwomen, *so extraordinary*, that I must confess I could not believe them; in truth, I was infinitely *more puzzled by what I heard, than by the simple evidence of my own eyes.*"

" I feel it right to state that, up to the period of my arrival at Oughterard, I had not, in Ireland, excepting in the police-cell in Dublin, seen one drunken person, either male or female."

" The devotional expressions of the lower class of Irish, and *the meekness and resignation with which they bear misfortune or affliction, struck me very forcibly.* . . . A PROTESTANT CLERGYMAN OF GREAT EXPERIENCE TOLD ME THAT, IN ALL HIS INTERCOURSE WITH IRISH CATHOLICS, HE HAD NEVER MET WITH AN INFIDEL."

" Why," said I to myself, as I finally closed the note-book of my little tour; " why, for so long a period, have the inhabitants of Ireland been centrifugally ejected from their country, as if its lovely, verdant surface were a land blasted by pestilence or as if its VIRTUOUS AND INTELLIGENT

PEASANTRY were malefactors, who had been sen-
tenced to transportation?"—Sir F. B. Head's
Fortnight in Ireland.

"Happy would it be if all who read the Scriptures more than this unnoticed woman," a poor old
Irishwoman, "would practice its precepts as well."

"*If the professed Christian, with the Bible in his
hand, do not know his duty towards the stranger,
then let him ' tie a string' around that Bible, and go
into some mountain cabin where the Bible has never
been*, AND THERE TAKE A LESSON.

"Does this look like idleness! Many a poor
widow have I seen, with some little son or daughter,
spreading her manure by moonlight, over her
scanty patch of ground; or before the rising of the
sun, going out with her whisp about her forehead,
and basket to her back, to gather her turf or potatoes."

"—— Yet the story of Calvary was well understood, *and they made a better application of the
Scriptures they did know, than do many who read
them daily.*"——"In no place did they appear dark
on the subject of Christ's death and sufferings."
Note, p. 296.

"—— Lamentable as it is, the lower class of
Protestants, wherever I have met them in Ireland,
are more ignorant of their religion than the same
class among the Catholics."

"The next day we visited a school of the nuns

Here were more than three hundred of the poor taught in the most thorough manner. Their lessons in grammar, geography and history, would do honor to any school, and their needlework was of the highest order.'

"—— I blessed the Father of all mercies that he had left in one island of the sea, a people who STILL RETAIN THE SIMPLE LIFE AND SIMPLE MANNER OF PATRIARCHAL DAYS."

"1 heard of Connemara, that it had been a custom from time immemorial, that if a stranger is not welcomed into a cabin at night-fall, or leaves it in a storm, the cabin-holder is immediately called upon to inquire into the reason; and if it appears that it is inhospitality, that family is set up as a mark of contempt to its neighbors."

"I asked the boy to read; he did so intelligibly, and answered every question from the second of Matthew, respecting the birth of the Saviour, correctly. . . . . . HE WAS READY IN THE SCRIPTURES, THOUGH HE HAD BEEN TRAINED IN THE CATHOLIC CHURCH."

"*Had my reception among the higher and middle ranks* (that is to say, the Protestants) *been as Christian-like and as civil as among the* (Catholic) *poor, it would have been one monotonous tissue, which might have spread a false coloring before my eyes, so that her* (Ireland's) *true character would have been hidder.*" (That is to say, had the writer only moved amongst the Catholic poor of Ireland, she would have been saved the cold inhospitality and haughty contempt and injurious suspicions which she in al most every instance experienced from the Protestant rich.)

"To the Roman Catholics, both duty and inclination require that I should acknowledge a deep debt of gratitude. They have opened the doors of convents, of schools, of mansions, and cabins, without demanding letters, or distrusting those that were presented. They have sheltered me from storm and tempest; they have warmed and fed me without fee or reward, *when my Protestant brethren and sisters frowned me away.* God will remember this, and I will remember it."

"The teacher observed that the Bible was daily read; ' and I find the children of the Catholics much more ready in the Scriptures than the Protestants, and make me much less trouble in getting their lessons. I cannot account for the fact, but so it is.' The circumstance is easily explained. THE SCRIPTURE WHICH IS EXPOUNDED TO THEM BY THEIR SPIRITUAL GUIDES, IS IMPRESSED AS BEING OF THE MOST AWFUL IMPORTANCE, AND ITS CONSEQUENCES OF THE MOST WEIGHTY IMPORT; AND WHEN THEY GET ACCESS TO THIS TESTIMONY OF GOD, THEY ARE PREPARED TO TREAT IT AS SUCH. THE PROTESTANT CHILD RELISHES IT NO BETTER THAN A STALE PIECE OF BREAD AND BUTTER, WHICH HE IS OFTEN FORCED TO

EAT AS A PUNISHMENT, WHEN HIS STOMACH IS AL-
READY SATIATED. AN INTELLIGENT GENTLEMAN FROM
DUBLIN REMARKED, THAT HE WAS WHIPPED THROUGH
THE BIBLE BY A PROTESTANT UNCLE WHEN A CHILD,
AND HAD HATED IT EVER SINCE." —MRS. NICHOLSON'S
Ireland's Welcome to the Stranger.

"They were Protestants. . . . But sorry am I to
say, that in no family had I heard so much pro-
fanity, both from mother and children. I would
not expose it, ... but such sins should be rebuked
before all.'—*Ibid.*

"Many favorable opportunities presented, to become acquainted with the effects of the famine
upon the Romish priests. . . . . They had two
drawbacks which the Protestants in general had not.
First, a great proportion of them are quite poor ;
and second, they, in the first season of the famine,
were not intrusted with grants, as the Protestants
were. . . . . One Protestant clergyman informed
me, that so much confidence had he in the integrity
of the Catholic priest in his parish, that when he
had a large grant sent to him, he offered as much
of it to the priest as he could distribute, knowing,
he added, that it would be done with the greatest
promptitude and fidelity. No ministers of religion
in the world know as much of their people as do
the Catholics, not *one* of their flock is forgotten,
scarcely by name, *however poor or degraded*, and
consequently, when the famine came, they had no

to search out the poor, they knew the identical
cabin in which every starving one was lying, and
. . . . were in a condition to act most effectually.'

"To do these poor priests justice, they have la-
bored long and hard since the famine, and have
suffered intensely. They have the most trying
difficulties to encounter, without the least remune-
ration. . . . . IN THE FAMINE, NIGHT AND DAY,
THEIR SERVICES WERE REQUISITE, NO FEVERS NOR
LOATHSOME DENS, NOR EVEN CAVES COULD EXONERATE
THEM; THEY MUST GO WHENEVER CALLED, AND THIS
WITHOUT ANY REMUNERATION."—MRS. NICHOLSON'
Annals of the Famine in Ireland.

PROSELYTISM.

"It requires the Irish language to provide suitable words for a suitable description of the spirit
which is manifested in some parts to proselytize,
by bribery, the obstinate Romans (Catholics) to
the Church which has been an instrument of oppression for centuries. The English language is
too meagre to delineate it in the true light. Rice,
Indian meal, and black bread would, if they had
tongues, tell sad and ludicrous tales. The artless
children too, who had not become adepts in deceit,
would and did sometimes by chance tell the story,
in short and pithy style. It was a practice by
some of the zealots of this class to open a school
or schools, and invite those children who were in
deep want to attend and instruction, clothes, and

food should be given, on the simple terms of read
ing the Scriptures and attending the church. The
Church catechism must be rehearsed as a substitute
for the Romish. . . . . . The children flocked by
scores and even hundreds; they were dying with
hunger, and by going to these places they could
'keep the life in them,' and that was what they
most needed; they could go on the principle, '*if
thou hast faith, have it to thyself before God*,' and
when the hunger was appeased, (they could go back
again to their own religion.) When such children
were interrogated, the answer would be, 'We are
going back again to our own chapel, or our own religion, when the stirabout times are over;' or 'when
the potatoes come again.'—'But you are saying
these prayers and learning this catechism.'—'We
shan't say the prayers when we go back—we'll say
our own then,' &c. Now the more experienced father or mother would not have said this to a stran
ger, and such might have passed for a true convert,
while receiving 'the stirabout.'—*Ibid*, pp. 300–301.

"The army is required to show its warlike power
in defence of the missionaries stationed there, being
called out to display their banners when any new
converts are to be added to the Protestant ranks
from the Romish Church. An instance of this was
related by a coast-guard officer, stationed in the
town of Dingle. Some five or six years ago, a
half dozen or more of the Romans had concluded

to unite with the Protestant mission established
there, and the Sabbath that the union was to take
place in the Church, the soldiery were called out to
march under arms, to protect this little band from
the fearful persecutions that awaited them on their
way thither. The coast-guard officer was summoned
to be in readiness *cap-a-pie* for battle, if battle should
be necessary; he remonstrated—he was a Methodist
by profession, and though his occupation was something warlike, yet he did not see any need of carnal
weapons in building up a spiritual Church; but he
was under government pay, and must do government work. He accordingly obeyed, and, to use
his own words substantially: ' We marched in
battle array, with gun and bayonet, over a handful
of peasantry—a spectacle to angels of our trust in a
Crucified Christ, and the ridicule and gratification of
priests and their flocks, who had discernment sufficient to see that, with all the boasted pretensions
of a purer faith and better object of worship (!)
both were not enough to shield our heads against
a handful of turf which might have been thrown
by some ragged urchin, with the shout of " turn-
coat" or " souper," as this was the bribe which the
Romanists said was used to turn the poor to the
Church; and though this was before the potatoe
famine, yet the virtues of soup were well known
then in cases of hungry stomachs, and the Dingle
mission had one in boiling order for all who came

to their prayers.'   The coast-guard continues
'We went safely to the Church, and the next mis
sion paper, to my surprise and mortification, told
a pitying world that so great were the persecutions
in Dingle, that the believing converts could not go
to the house of God to profess their faith in Him
without calling out the soldiery to protect them.' "
—*Ibid*, pp. 303, 304.

" The Roman Catholics are peculiarly distinct in
one noble practice, from all other professed Chris
tians we meet.   They will not in the least gape
after, nor succumb to any man's religion, because
he is great and honorable ;—where their religious
faith is concerned, they call no man master."—*Ibid*,
p. 314.

" The old hackneyed story of Popery in Ireland has been so turned and twisted that every
side has been seen—nothing new can be said
against it.   There it stands . . . . the same in
essence, as when Queen Elizabeth put her anathemas
forth against its creeds and practice ; and, with all
her errors (!) she maintains a few principles and
practices *which it would be well for her more Bible
neighbors to imitate.* HER GREAT ONES ARE MORE
ACCESSIBLE ; THE POOR OF THEIR OWN CLASS, OR OF
ANY OTHER, ARE NOT KEPT AT SUCH AN AWFUL DIS-
TANCE ; THE STRANGER IS SELDOM FROWNED COLDLY
FROM THEIR DOOR ; TO THEM THERE APPEARS TO BE A
SACREDNESS IN THE VERY WORD WITH WHICH THEY

WOULD NOT TRIFLE; THE QUESTION IS NOT, IS I E OR
SHE 'RESPECTABLE,' BUT A STRANGER; IF SO, THEN
HOSPITALITY MUST BE USED WITHOUT GRUDGING. In
the mountains and sea-coast parts, it has ever been
the custom to set the cabin door open at night,
and keep up a fire on the hearth, that the way-far
ing man and the lone stranger, should he be
benighted, could see by the light that there is wel
come for him; and if they have but one bed. the
family get up and give it to the stranger, sitting
up, and having the fire kept bright through the
night.   This has been done for me, without knowing
or asking whether I was Turk or Christian; and were
I again to walk over that country, or be out at
nightfall in storm or peril, as has been my lot, and
come in sight of two castle-towers, one a Roman,
and the other a Protestant owner; *and were the
former a mile beyond, my difficult way would be made
to that, knowing* that when the porter should tell
the master a stranger was at the gate, he would
say : ' Welcome the stranger in for the night, or
from the storm.' "—*Ibid*, p. 328.

   " THE CATHOLICS ARE MUCH MORE HUMBLE IN
THEIR DEMEANOR, AND CERTAINLY MUCH MORE HOSPI
TABLE AND OBLIGING IN ALL RESPECTS, AS A PEOPLE
THEY ARE MORE SELF-DENYING, WILL SACRIFICE THEIR
JWN COMFORTS FOR THE AFFLICTED, M(RE READILY
WILL THEY ATTEND THEIR PLACES OF WORSHIP, CLOTH
ED OR UNCLOTHED, AND BEGGARS TAKE AS HIGH A

PLACE OFTEN IN THE CHAPEL AS THE RICH MAN."—
Ibid, p. 329.

This, then, reader! is a picture of the Irish
people as they are. We here learn from good
Protestant authority—for I have quoted no other—
that they are, take them all in all, a nation of hum
ble, practical Christians ; chaste, modest, patient,
kind and hospitable—*enduring all things*——ay!
more than ever nation bore—yet enduring with a
cheerfulness, a resignation that could only have
their source in the purest and holiest spirit of
Christianity. Mrs. Nicholson, a member of the
Presbyterian Church, in her two remarkable works
on Ireland, gives innumerable instances of the pa
tient endurance of the poor starving people, and
their cheerful resignation to the will of God ;—she
makes grateful mention of their spontaneous kindness and their heroic self-denial in practising the
Christian duty of hospitality. She unhesitatingly
admits that in these respects the Protestants—
ministers and all—were not to be compared to the
Catholics ;—she gives a graphic picture of the cold,
selfish hypocrisy of the Achill ministers, their
utter want of even common kindness, and their
injurious treatment of the well-meaning, though
somewhat fanatical *stranger*. She relates other
instances of the treatment she received at the
eands of rich Protestant philanthropists and Pro
selytizers—for the most part, synonymous terms—

which are admirably calculated to show the differ
ence between Charity and Philanthropy—the
former beautifully illustrated in the poor, humble,
unpretending Catholic, and the latter in the rich
Protestant patrons of the New Reformation. And
yet such is the force of fanaticism that this very
woman still bewails the influence of *Romanism*,
and sighs for the advent of a purer religion—looks
forward to the Scriptural enlightenment of the
people as the grand means of improving their con·
dition—that is to say, she would have them become rich and comfortable in this Scriptural religion of hers, at the expense of their Christian virtues
and endearing qualities. She would have them
" *go to Christ*," when it is clear as the noonday
sun—even from her own showing—that the spirit
of Christ dwells with them—if it did not, how
would they suffer hunger and cold and nakedness,
and behold their nearest and dearest dying of starvation, and yet *bless God*, as did the holy man Job
under *his* afflictions. Sir Francis Bond Head, who
is anything but favorable to the Catholic religion,
says that it is quite extraordinary to hear these
poor people praising and blessing God in the midst
of all their sufferings. Why, then, would he and
his rob them of that old, firm faith, and that Catholic devotion which has cheered and consoled their
fathers, and still does the same for them? What
would the proselytizers have? Do not the Catholic

people of Ireland love God and hope in him?—"In no place did they appear in the dark on the subject of Christ's death and suffering."—"A Protestant clergyman of great experience said that in all his intercourse with Irish Catholics he had *never* met an infidel."—" They are taught to regard the Scriptures with greater reverence, and as being of awful importance."—"They are a nation of practical philanthropists." Their women are admitted to have an innate modesty, and to be more chaste than any other women known to the Protestant writer—"their great ones are more accessible"—"they are more humble in their demeanor." What, then, I repeat, would the proselytizers have?—Will they dare to maintain the palpable absurdity that the religion of these people is *not* the religion of Christ?—or that the religion of the Achill ministers, and the hard-hearted, proud, self-righteous philanthropists *is?* Even they, it would seem, could scarcely maintain such a barefaced falsehood.

With regard to the old, stale calumny that the Catholic religion has the effect of stultifying the mind and freezing " the genial current of the soul," I might quote innumerable Protestant authors to prove the contrary. I shall only give one quotation on the subject. It is Mrs. Nicholson who again speaks. Hear her describe a Catholic lady and her family:—"The piano and the harp, the ancient boast of Ireland's better days, were there

and the lady, *who had been educated in a convent,*
knew well how to touch the heart by her melody.
Her two little daughters, who were but children,
did honor to her who had trained them with a skil
ful hand. Never had I seen high birth, beauty,
AND NOBLE INTELLECTUAL ATTAINMENTS MORE HAP
PILY BLENDED WITH A MEEK AND QUIET SPIRIT THAN
IN THIS ACCOMPLISHED WOMAN. Though she was a
Roman Catholic, yet *the higher class of Protestants
were anxious to place their daughters under her care."*
Mrs. Nicholson's surprise only goes to prove that
she knew as little of the real workings of the
Catholic religion as she did of Catholic ladies. Of
all the impudent fictions ever palmed upon the cre-
dulous, that of Catholicity being incompatible with,
or inimical to the cultivation of the mind, or the
progress of art and science, is the most audacious,
because the most unfounded. How amusing is it—
yet withal provoking, to hear the half-educated,
perhaps wholly illiterate Scripture-reader, holding
forth to the astonished natives of some wild Connemara glen on "the darkness of Popery"—"the
grievous bondage wherein Popery holds the human
mind"—"the glorious light and liberty enjoyed by
Protestants," &c., &c. How little does the poor
drivelling ranter himself know of "Popery"!—
how little does he think that the greatest, best, and
most enlightened men whom the world has ever
seen have been and are Roman Catholics—that the

face of Europe is covered with the immortal crea-
tions of *Popish* genius—that the stately cathedrals
erected to the glory of God in Catholic times are
still the admiration of the world—little dreams he
of what Michael Angelo, the greatest painter who
has yet lived—Rubens—Rembrandt—Canova—Ti-
tian—Claude Lorraine—Carlo Dolchi—Guido—
Tasso—Dante—Pope—Dryden—all Catholics,
have done for the arts and human letters—nor
what Catholic missionaries and Catholic martyrs
have done for religion.

St. Francis Xavier, St. Francis de Sales, St. Ignatius Loyola, Fenelon, Bossuet, More, and Fisher
are utterly ignored, and so is the grand truth that
the great lights through whose agency God has il-
lumined the earth, were and are, for the most part,
Catholics—that the great universities of Europe
were almost all founded by Catholics—that the
Constitution of which Protestant England is so
proud, is principally the work of Catholic kings
and nobles in the old "ages of faith"—that the no-
blest actions on record were achieved by Catholics
—that Wallace, and Tell, and Hofer were Catholic
to the heart's core, though Protestants—the
ephemeral offspring of latter ages—do *modestly*
descant upon "the slavish spirit of Catholics"—
"the debasing influence of Rome," &c., &c. Oh!
for a tongue to make those poor Connemara moun-
taineers hurl back the base calumnies heaped upon

their fa. :h—that faith which is only known to them
as *the true religion*—the consoler of their affliction—
the strength of their weakness—the hope of their
sufferings—the light of their darksome path : they
know nothing, poor, simple Christians, of the radiant halo that encircles the brow of that divine
religion :

> "For knowledge, to their eyes, her ample page,
> Rich with the spoils of time, hath ne'er unroll'd."

But though they cannot look back through the
pages of history, they *can* through the traditions
of their fathers; these tell them of a period when
the land of Ireland was all Catholic; when the heretic
or the stranger found not his way into their Alpine
regions—when peace and plenty prevailed, and
men and women lived for heaven, content with
whatever little God might have given them here
on earth, and willing to share it with those who
had still less, and so they lived happy and died
well. These are the traditions handed down
amongst the Catholic people of Ireland, and they
are as a wall of adamant guarding the nation's
faith. The proselytizer may spend his thousands
and thousands of English gold, providing Bibles,
and tracts, and "stirabout," and soup—he may
flatter himself and the people who fill his pockets
that he is doing wonders amongst the Irish papists—he may succeed to a certain extent, while

famine continues to desolate the land—(there are
always to be found, even amongst 'the virtuous
and intelligent peasantry' of Ireland, some few
scape-goats who go out into the desert, bearing, I
trust, the sins of the people)—but when once
the scourge has passed away, and 'plenty
smiles again on the land,' then the proselytizer, whether hypocrite or fanatic, shall see the
whole castle of his hopes topple to the ground,
and his beautiful *Fata Morgana* melt into air. He
will find out the truth of what a certain car-driver
said to Sir Francis B. Head (though I must take
this opportunity of protesting against that gentleman's attempts at Irish phraseology or pronuncia-
tion—both are entirely at fault) :

"A number of workmen," says Sir Francis,
"were busily erecting a large, substantial stone
Protestant Church, with Gothic windows.

"'Thart's,' said the driver, as he pointed to it
with his whip, 'for what we ca' "Joompers;" but
if the pittaturs would return, they'd a' come back.
They would, indade, your arn'r."—p. 153.

And who can doubt that the man spoke the
truth? Does not every day's experience show
the poor Jumpers or Soupers (as they are deri-
sively called) returning to the old religion, when
once the pressure of famine is past? When they
get money from abroad, or permanent employment
at home, is not "their first race," as they would

say themselves, " to the priest," and their first act
to become reconciled to that holy Church, which
their temporary apostacy has made all the more
dear to their heart ?   But above all, when death
begins to approach—if time be given them—they
almost invariably cry out for " the priest," and
regard the public scandal they have given as the
greatest, the most fearful of crimes.   If any proof
were wanted to show the true character of this
persevering attack on the ancient faith of Ireland,
it would be found in the savage fury of the proselytizers when these poor people escape from their
clutches, and return to the Church.   Thus we see
them at one time bringing a suit against a poor
man, for the clothes they had given him when he
went to their Church—said clothes being the bribe
meant to buy up his faith—at other times we see
them suffering poor widows and other desolate
creatures *to die of hunger*, because they would not
take relief at the expense of their hopes of heaven !
Again we see them taking back, with the most
unfeeling harshness, whatever they had given, because the poor recipients of their bounty had at
last acted on the dictates of conscience, and sought
refuge once more in what they knew and felt was
the ark of safety.   One of the latest instances of
this kind is especially deserving of attention.   A
poor man had been forced by the pangs of l unger
" to conform ;" the proselytizers gave him   com

fortable cottage, " together with all the adjuncts ;'
he remained for several years (to all appearance)
" a good man and true"—that is to say, a Jumper
but at last, being taken sick, he sent for the priest,
whereupon the *Bible-Christians* came in strength
to dissuade him from returning Rome-wards (and
home-wards); not being able to succeed (for the
fear of death was before the sick man's eyes), what
does the reader suppose they did ? why they carried the sick man out, placed him on the road, and
then tore the roof off the house, lest he or his
might find shelter there again.   Never, in the annals
of the world, has there been so cruel a " sham," so
' great a delusion" practised on mankind, as this of
the Protestant attempts to *convert* Catholics, and
above all, the Catholics of Ireland.   The proselytizers find the Irish " Papists" such as I have shown
them to be, on unquestionable Protestant authority ;
they find them pious, chaste, humble, patient,
temperate, kind, generous, hospitable, bearing all
things with resignation for God's sake; they would
make them what ?—why, as unchaste and immoral
as the Protestant nations around them, where
thousands, millions of the people know not God or
our Lord Jesus Christ, even in name; where all
manner of wickedness abounds, and the things of
earth entirely supersede the things of heaven.
They come to them, in their hypocritical kindness,
with the open Bible in their hand, telling them to

take and read," just as though poor, simple, illi
carate creatures like them are fit to fathom the
sublime profundity of Holy Writ, which even the
most learned of the Doctors of the Church approach
with reverence and awe. Why, the bare idea is
preposterous, well nigh blasphemous.

In conclusion I will quote, for the benefit of the
Protestant reader, those memorable words of the
late Richard Lalor Shiel, Ireland's great orator,
himself a faithful son of the Most Holy Church of
Christ :

" The Catholic religion, indigenous to the soil of
Ireland, has struck its roots far and deep in the
hearts and affections of her people ; it grows
beneath the axe, and opens with the blast ; whilst
the Protestant creed, though preserved in a magnificent conservatory, at a prodigious cost, pines
away like a sickly exotic, to which no natural
vitality can be imparted."

It would be well if the Irish proselytizers and
their supporters made a deep and earnest study of
this text ; they would, perhaps, become both wiser
and better men, and might save themselves a
world of trouble, and useless trouble, too, for,
with the blessing of God, the children of St. Patrick shall continue to be as they have ever been,
immovably attached to the chair of Peter, and
guided by the old lamp of faith.

THE END.

www.ingramcontent.com/pod-product-compliance
Lightning Source LLC
Chambersburg PA
CBHW021345210326
41599CB00011B/753